NUCLEAR MAGNETIC RELAXATION

BY

Dr. N. BLOEMBERGEN

Springer-Science+Business Media, B.V. 1948

ISBN 978-94-017-5723-2   ISBN 978-94-017-6073-7 (eBook)
DOI 10.1007/978-94-017-6073-7
Softcover reprint of the hardcover 1st edition 1948

Aan mijn Ouders.

The work described in this thesis was carried out at Harvard University, Cambridge, Massachusetts, in the years 1946 and 1947, under the supervision of professor P u r c e l l.

Most of the information presented here may be found in a paper to be published jointly by P u r c e l l, P o u n d and the present author in the „Physical Review". A brief account has already appeared in „Nature".

I am greatly indebted to the members of the Physics Department of Harvard University for increasing my knowledge of physics, but also for their helpfulness in other respects. Especially I would like to mention you, professor P u r c e l l. The years of cooperation and friendship with you and P o u n d I shall not forget. It is unnecessary to say how greatly I benefited in numerous discussions from the knowledge of both of you, who performed the first successful experiment of nuclear magnetic resonance, together with T o r r e y. Several discussions with professor v a n V l e c k and a stimulating course of professor S c h w i n g e r have also contributed to this thesis. The work was supported by a grant from the Research Corporation.

Mijn verblijf in Harvard zou echter minder resultaat hebben gehad, indien ik niet reeds in Utrecht de eerste schreden op het pad der experimentele natuurkunde gezet had onder Uw leiding, Hooggeleerde M i l a t z, en indien niet U, Hooggeleerde R o s e n f e l d, mij de beginselen van de theoretische physica had geleerd.

Hooggeleerde G o r t e r, dat U, wien het onderwerp van dit proefschrift zo na aan het hart ligt, mijn promotor wilt zijn, stemt mij tot grote voldoening. Dat ik U niet als eerste noem, is slechts daaraan te wijten, dat ik U niet vroeg genoeg heb leren kennen. Ik dank U voor Uw waardevolle critiek gedurende de laatste stadia van dit proefschrift.

Nog steeds ben ik dankbaar, dat het Utrechts Stedelijk Gymnasuim aan het begin van mijn studie heeft gestaan.

CONTENTS

8

CHAPTER I.

INTRODUCTION.

1. 1. *Nuclear spin and moments.*

The fundamental properties of a nucleus can be listed in the following way (B 1):

Mechanical properties
{
  1. Mass.
  2. Size.
  3. Binding energy.
  4. Spin.
  5. Statistics.
}

Electrical properties
{
  6. Charge.
  7. Magnetic dipole moment.
  8. Electric quadrupole moment.
}

In this thesis we shall only be concerned with properties 4, 7 and 8 of the list.

In 1924 Pauli (P 2) suggested that the hyperfine structure in atomic spectra might be explained by a small magnetic moment of the nucleus. The interaction of this magnetic dipole with the motion of the electrons would produce a hyperfine multiplet in a similar way as a multiplet is produced by interaction of the intrinsic magnetic moment of the electron with the orbital motion. The introduction of the electronic spin by Uhlenbeck and Goudsmit in 1925 made it possible to explain many hitherto mysterious details of the spectra. It appeared appropriate to connect the magnetic moment of the nuclei also with rotating charges and to attribute to the nucleus a mechanical spin. The evidence for the nuclear spin and nuclear magnetic moment is now manifold, and the concept of the spinning nucleus must be considered to be as a well founded as that of the spinning electron. The nuclear spin can be determined from the following phenomena :

1. Alternating intensities in band spectra.
2. Intensities, interval-rule and Zeeman-splitting in hyperfine multiplets.
3. Polarisation of resonance radiation.
4. Magnetic deflection in atomic and molecular beams (Stern, Rabi).

5. Total intensity of resonance absorption in nuclear paramagnetism.

6. Specific heats of (ortho- and para-) hydrogen and deuterium.

7. Scattering of identical nuclei.

The magnitude of the magnetic moment can be determined from the following experiments:

1. Splitting and Zeeman-splitting in hyperfine multiplets.

2. Magnetic deflection in molecular beams.

3. Magnetic resonance in molecular beams.

4. Nuclear paramagnetism.

5. Resonance absorbtion and dispersion in nuclear paramagnetism.

6. Ortho — para — conversion of hydrogen.

In 1935 Schüler and Schmidt (S 2) observed deviations from the interval rule in the hyperfine multiplets of Europium, which could be explained by assuming another type of interaction between the nucleus and the surrounding electrons, by means of a nuclear electric quadrupole moment. The experimental evidence for the existence of these moments has since grown, and is produced by:

1. Hyperfine spectroscopy.

2. Magnetic deflection of molecular beams (zero-moment method).

3. Fine structure of the magnetic resonance line in molecular beams.

4. Fine structure in microwave spectra.

5. Relaxation phenomena in nuclear paramagnetism,

6. Ortho — para — conversion of deuterium.

So far an influence of moments of higher order, the magnetic octupole or electric sedecipole, has not been discovered. It is, of course, outside the scope of the thesis to give even a short description of all these methods. Only the effects of nuclear paramagnetism are our topic, but we shall find opportunity to give a very brief discussion of the molecular beam method in section 1.3 and 1.5 of this introductory chapter, since these experiments are closely related to our subject. For the other fields we let follow some references, which will introduce the reader to the literature.

Kopfermann (K 5) gives a discussion of all effects which were known up to 1939 with complete references. Especially the hyperfine spectra are discussed extensively. An account of the properties of ortho — and para —, light and heavy hydrogen has been given by Farkas (F 1), the influence of the quadrupole moment of the deuteron on the ortho — para — conversion was treated by Casimir (C 1).

A field which has only recently become accessible by the development of radar techniques during the last war is microwave spectroscopy. The influence of the existence of quadrupole moments on these spectra is discussed by several authors (B 14, C 3, D 1, T 1).

The scattering of identical nuclei, especially the $p-p$ and $\alpha-\alpha$ scattering, has been discussed in numerous papers (B 3, G 1, M 4).

The nuclear spins and moments are of great interest in the theory of the constitution of the nuclei. The reader is referred to the detailed theory of the deuteron, and of H^3 and He^3 (G 2, L 3, S 1, S 4). The numerical values of the moments of these particles are explained in terms of those of the elementary particles, the neutron and proton. Some rather crude models have been built for heavier nuclei. Spin and magnetic moment also play an important role in nuclear reactions, of which we may mention the spin selection rule in β-decay, the neutron-proton scattering and the photo-magnetic desintegration of the deuteron. However in general the situation is such that the most accurate information about nuclear spins and moments is obtained from the methods mentioned in the beginning, which usually are not considered to belong to the field of nuclear physics. The results are useful to further development of nuclear theory and understanding of nuclear reactions, rather than that these latter processes yield the values of spin and moment. The case of the triton (H^3) offers an interesting illustration (A 3, B 5, S 1).

1.2. *Mathematical introduction of the nuclear spin.*

The spin is a purely quantummechanical concept. The spin angular momentum has similar properties as the orbital angular momentum. They behave in the same way under a rotation of the coordinate system and have the same commutation rules. The rules for quantisation and composition of these momenta are set forth in detail in textbooks on quantummechanics (C 4, K 6). Here we shall briefly summarize some results. The square of spin angular momentum $\vec{I}$ has the eigenvalues $I(I+1)\hbar^2$. While an orbital quantumnumber can only assume integer values, for the spin quantumnumber I also half integer values are allowed. It has been found that the spin of neutrons and protons is $1/2$, just as for the electron. The nuclear spin is generally composed of the spins of these elementary particles and their angular momenta in the nucleus. Therefore the nuclear spin I must be an integer or half integer according to whether the number of these so called nucleons is even or odd. On somewhat obscure grounds it is believed that all nuclei possessing an even number of protons and an even number of neutrons have spin zero. Experimentally this is confirmed for the lighter elements from band spectra. The experimental information for heavier isotopes only indicates that the magnetic moment, if at all present, is very small, and thus not in disagreement with a spin zero.

The nuclear spin I will, under all circumstances considered in this thesis, be a constant of the motion, since the energies involved in the processes described will never be large enough to produce transitions of the nucleus to an excited state. Only reorientations will occur. The matrices of the three components of the spin angular momentum operator transformed to the representation, in which the z-component of the operator is diagonalised, are:

$$(m \mid I_x + i\,I_y \mid m-1) = \sqrt{(I+m)(I-m+1)} \tag{1.1}$$

$$(m \mid I_x - i\,I_y \mid m+1) = \sqrt{(I-m)(I+m+1)} \tag{1.2}$$

$$(m \mid I_z \mid m) = m \tag{1.3}$$

All other elements vanish.

The z-component of the spin operator has $2\,I+1$ eigenvalues $I,\ I-1,\ \ldots\ ,-I$; m_I is called the magnetic spin quantumnumber. Not only I, but also the total electronic quantumnumber J of the atom or molecules under consideration in this book will be a constant of the motion. Processes in which the electronic state might change, will not occur. Now the angular momenta $\vec{I}$ and $\vec{J}$ combine to a resultant $\vec{F}$ in the same way, as in the case of Russell-Saunders coupling the total electronic spin $\vec{S}$ and total electronic angular momentum $\vec{L}$ combine to $\vec{J}$. The component of $\vec{F}$ on any preferred axis, which is usually taken in the z-direction, can assume the values $F,\ F-1,\ \ldots\ ,-F$. With the angular momentum vector $\hbar\,\vec{I}$ is connected the magnetic moment vector $\vec{\mu}$. Since the two vectors behave in the same way under a rotation of the coordinate system we must have

$$\vec{\mu} = \gamma\hbar\,\vec{I} \tag{1.4}$$

where $\gamma = \dfrac{\text{magnetic moment}}{\text{impuls momentum}}$ is the magnetogyric ratio, which can be either positive or negative, $\gamma\,\hbar\,I$ is the maximum eigenvalue of the z-component of the magnetic moment operator and is often called the nuclear magnetic moment. $I\,\hbar$ is the maximum value of the z-component of the spin momentum vector.

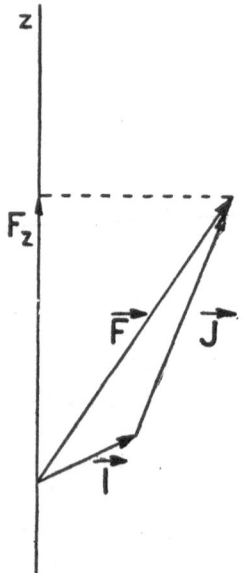

Figure 1. 1.

Coupling of the electronic and nuclear momenta in the absence of an external field.

For one free electron we have in analogy to (1. 4)

$$\vec{\mu}_{el} = \gamma_{el}\, \hbar\, \vec{S}_{el}$$

The intrinsic magnetic moment of an electron (maximum z-component) is one Bohr magneton $\beta = e\,\hbar/2\,m\,c$ and $\gamma_{el} = e/mc$ is twice the classical ratio, which holds for the orbital motion. One must not confuse γ with the dimensionless Landé-factor $g = \dfrac{\text{magnetic moment expressed in Bohr magnetons}}{\text{impuls momentum expressed in units } \hbar}$. For one free electron we have experimentally $g = 2$, which is also a result from Dirac's relativistic theory of the electron. For one or more bound electrons the total magnetic moment has in general not the same direction as the total angular momentum $\vec{J}$, as the magnetogyric ratios for spin and orbital motion are different. But the low frequency components of the magnetic moment operator, in which we shall only be interestered, can still be represented by

$$\vec{\mu} = g\,\beta\,\vec{J}$$

If the proton obeyed the Dirac equations, as the electron does, the magnetic moment of the proton would be one nuclear magneton

$$\beta_{nuc} = e\,\hbar/2\,Mc$$

Experimentally one finds that this is not true. Nevertheless the nuclear magneton gives the order of magnitude of the nuclear magnetic moments.

We list the values of spins and moments of the isotopes, used in the experiments, described in this thesis, in the following

TABLE

Nucleus	Spin	Magnetic moment	Electric quadrupole moment	Magnetogyric ratio
H^1	$I = 1/2$	$\mu = 2.7896$	$Q = 0$	$\gamma = 2.673 \times 10^4$
H^2	1	0.85647	2.73×10^{-27}	0.4103×10^4
Li^7	$3/2$	3.2535	?	1.039×10^4
F^{19}	$1/2$	2.625	0	2.517×10^4

The impulsmomentum is expressed in units $\hbar = 1.054 \times 10^{-27}$ erg sec.
The magnetic moment $\mu = \gamma\,\hbar\,I$ is expressed in units $\beta_{nuc} = 5.049 \times 10^{-24}$ erg oersted^{-1} .
The magnetogyric ratio is expressed in oersted^{-1} sec^{-1} .
The quadrupole moment Q, defined in chapter 5, is expressed in cm^2.

The spin of about 90 isotopes is known, the magnetic moment of about 50, the quadrupole moment of about 10. For the heavier isotopes usually only spectroscopic data are available. From these the values of the moments can be obtained only, when the magnetic field and the gradient of the electric field at the nucleus are known.

Now just as a Paschen-Back effect can occur in the electronic Russell-Saunders coupling, it will be possible to change the quantisation of $\vec{J}, \vec{I}$ and $\vec{F}$ by means of an external magnetic field. As a matter of fact it will be much easier in the latter case to decouple $\vec{J}$ and $\vec{I}$, since the interaction energy between nucleus and electron system is much smaller than between electron spin and electron orbit. In spectroscopic language one would say: the hyperfine multiplets are narrower than the multiplets. The general form of the Hamiltonian for an atom or molecule in a constant magnetic field H_0 is:

$$H_{op} = A \, \gamma_I \, \hbar \, \vec{I}.\, \vec{J} + g \, \beta \vec{J}.\, \vec{H}_0 + \gamma_I \, \hbar \, \vec{I}.\, \vec{H}_0 \qquad (1.5)$$

$A \vec{J}$ represents the magnetic field at the position of the nucleus produced by the electronic motion. If the second or third term is large compared ot the first, a Paschen-Back effect occurs. Instead of the set of quantumnumbers I, J, F, m_F we have then the set I, J, m_I, m_J. Since the magnetic moment of the electrons is roughly 10^3 times larger than the nuclear moments, the second term is always larger than the third, unless the atom or ion is in an S-state, for which $J = 0$.

The ground state of molecules is usually a $^1\Sigma$ state. Then there is no contribution from the electron orbits and spins to the magnetic moment, but there is a small contribution from the rotation of the whole molecule. We can use the same formula (1. 5). where $\vec{J}$ now stands for the rotational angular momentum of the molecule. In this case the second and third term are of the same order of magnitude.

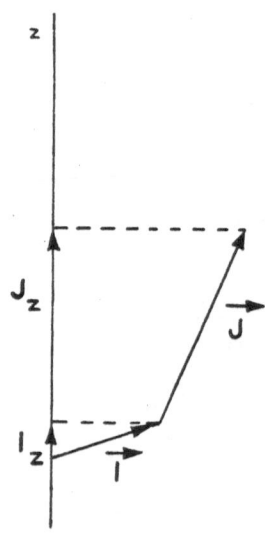

Figure 1. 2.
Paschen-Back effect; decoupling of electronic and nuclear momenta in a strong external magnetic field.

1. 3. *Atomic beam deviation method.*

In 1921 S t e r n and G e r l a c h let a beam

of silver atoms pass through an inhomogeneous magnetic field. A force $\vec{K} = (\mu.\ \text{grad})\ \vec{H}$ acted on them. The beam split in two parts. This could be explained by quantumtheory. The magnetic moment of the atoms with $J = \frac{1}{2}$ could only assume two positions with respect to the direction of the field, either parallel or antiparallel. The force acting in the cases would have opposite direction. The method is usually referred to as the atomic beam method, because usually only strong electronic moments produce a satisfactory deflection. But in 1933 Stern applied his method to hydrogen molecules (E 2, F 2). In parahydrogen the deflection is entirely caused by the rotational moment of the molecule, but in ortho-hydrogen an additional effect of the nuclear moments could be detected. It was then found that the proton moment is roughly 2.5 times the nuclear magneton. Rabi has shown how one can obtain information about the hyperfine splitting and nuclear spin in atomic beams in spite of the presence of the large electronic moments. The values of the magnetic field must be taken so low that the coupling between $\vec{J}$ and $\vec{I}$ is not destroyed, and the first and second term in (1.5) are of the same order. For a description of these beautiful experiments the reader is referred to Kopfermann and the original literature there mentioned.

1.4. *The resonance principle.*

Any system possessing an angular momentum $\vec{J}$ and a magnetic moment μ, which is placed in a magnetic field rotating about the z-axis, is able to reorient itself in this field. The theory was first given by Güttinger (G 7) and Majorana (M 1) and later with more generality by Rabi and Schwinger (R 1, S 3). A more detailed discussion of this phenomenon will be given in chapter 2. Here we shall merely indicate, in a rough manner, the nature of the process involved. For this purpose we first make use of the classical picture. Let H_0 be the constant z-component of the field and H_1 the components rotating in the xy-plane.

$$H_x = H_1 \cos \omega t$$
$$H_y = H_1 \sin \omega t \qquad (1.6)$$
$$H_z = H_0$$

We assume here $H_1 \ll H_0$. Ignoring H_1 for the moment, the magneto-mechanic system will classically precess around H_0 with the Larmor frequency

$$2 \pi \nu_0 = \omega_0 = \gamma H_0 \qquad (1.7)$$

The field H_1 will exert a torque

$$\vec{T} = [\vec{\mu} \times \vec{H_1}] \qquad (1.8)$$

This torque tends to change the angle between $\vec{\mu}$ and the z-axis. If, however, H_1 and the system rotate in opposite directions, or if they do not rotate with the same frequency, the torque will soon get out of phase and after a short time interval change its sign, so that the average effect over many Larmor periods will be small. If $\omega = \omega_0$, the polar angle will gradually increase; a reorientation takes place at resonance. In the first paragraph of chapter 2 the classical description is continued in more detail.

In quantummechanical language (B 9) the effect can be described as an „optical" transition between two energy levels. We suppose that the first two terms in the Hamiltonian (1. 5) can be omitted. For molecules in a $^1\Sigma$ state in a sufficiently strong magnetic field this is certainly allowed, and it is rigorously correct for atoms or ions in a ^{1}S-state.

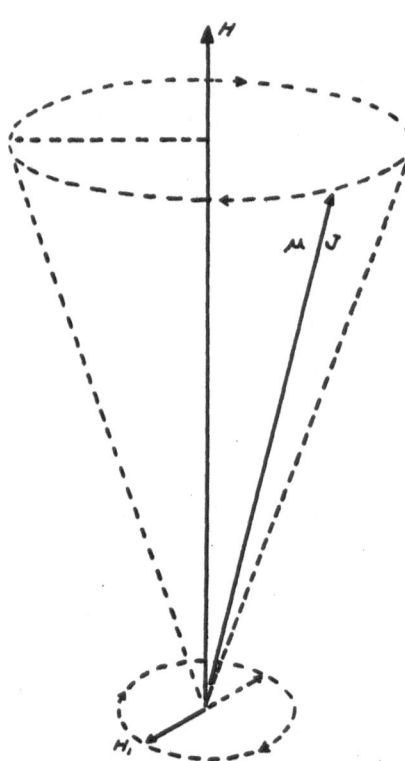

Figure 1. 3.

Simple vector representation of a magnetic gyroscope in a large fixed magnetic field H and a small rotating field H_1, according to K e l l o g g and M i l l m a n (Rev. Mod. Phys. 18, 325, 1946).

The Hamiltonian for the problem is

$$H_{op} = - \gamma \hbar \, \vec{I}_{op} . \vec{H} \qquad (1.9)$$

The rotating magnetic field can be represented by

$$\begin{aligned} H_x + i\,H_y &= H_1 \, e^{+i\omega t} \\ H_x - i\,H_y &= H_1 \, e^{-i\omega t} \\ H_z &= H_0 \end{aligned} \qquad (1.10)$$

Consider the magnetic field H_1 as a small perturbation. The unperturbed Hamiltonian

$$H^0_{op} = - \gamma H_0 I_z \qquad (1.11)$$

has $2\,I+1$ eigenvalues, corresponding to the $2\,I+1$ diagonal matrix elements of I_z in (1.3). The perturbation operator is

$$
\begin{aligned}
H^1_{op} &= - \gamma \hbar \, (H_x I_x + H_y I_y) \\
&= - {}^1\!/_2 \, \gamma \, \hbar \, H_1 \, \{(I_x - i I_y) e^{+i\omega t} + (I_x + i I_y) e^{-i\omega t}\}
\end{aligned}
\qquad (1.12)
$$

H^1_{op} has only non-diagonal elements, given by (1.1) and (1.2). In this problem one has to supply these two matrices with a Heisenberg time factor $e^{+i\omega_0 t}$ and $e^{-i\omega_0 t}$ respectively. The only effect of H^1_{op} is to produce transitions between adjacent levels. One has the selection rule $\triangle m_I = \pm 1$. Applying the usual first order perturbation theory one obtains for the probability to find the system in the state m' at time t, while at $t = 0$ it was in the state m:

$$w_{m \to m'} = \frac{1}{\hbar^2} \left| (m' | H^1 | m) \right|^2 \left| \frac{1 - e^{i\triangle\omega t}}{\triangle\omega} \right|^2 \qquad (1.13)$$

where

$$\triangle\omega = \omega - \omega_0$$

Substituting (1.12) and (1.2) into (1.13) one finds:

$$w_{m+1 \leftarrow m} = {}^1\!/_4 \, \gamma^2 \, H_1^2 \, (I - m)(I + m + 1) \sin^2 \frac{\triangle\omega t}{2} \Big/ \left(\frac{\triangle\omega}{2}\right)^2 \qquad (1.14)$$

This expression is very small, unless $\triangle\omega \approx 0$. Note that the resonance condition $\omega = \omega_0$ must also be satisfied to the sign in the complex phase factors $e^{\pm i\omega t}$. If the magnetogyric ratio is positive, H_1 must rotate counter clockwise, looking in the direction of H_0. Using the numerical value $\gamma \hbar I = 1.4 \times 10^{-23}$ E.M.U. for the moment of a proton, one finds that the resonance in a field of 6800 oersted occurs at 29 Mc/sec, that is, in the radio frequency range. If the radio frequency signal is spread out over a small frequency range containing the resonance frequency, and if we denote the average energy density stored in the rotating component of the magnetic field by $\int_0^\infty \varrho\,(\nu)\,d\nu = \dfrac{H_1^2}{8\,\pi}$ we have to integrate over $\triangle\omega$ in (1.14) and obtain in the familiar way a time proportional transition probability:

$$w_{m+1 \leftarrow m} = 2\,\pi\,\gamma^2\,(I - m)(I + m + 1)\,\varrho\,(\nu_0)\,t \qquad (1.15)$$

If we assume that the chance to be in the initial state m is equal for all m's, we can average (1.14) over m and obtain for the probability that any transition $\triangle m = +1$ is made

$$w_{\triangle m = +1} = \gamma^2 H_1^2\,{}^2/_3\,I\,(I+1)\,\sin^2 \frac{\triangle \omega\,t}{2} \Big/ (\triangle \omega)^2$$

and (1.15) goes over into;

$$w_{\triangle m = +1} = 2\,\pi\,\gamma^2\,\varrho\,(\nu_0)\,t\,{}^2/_3\,I\,(I+1) \qquad (1.16)$$

In the evaluation of the expression
$$\frac{1}{2I+1} \sum_{m=-I}^{+I} (I-m)(I+m+1)$$ the well known relations $\sum_{m=-I}^{+I} m^2 = {}^1/_3\,I\,(I+1)\,(2\,I+1)$ and $\sum_{m=-I}^{+I} m = 0$ have been used. The probability for a transition $\triangle m = -1$ is, of course, given by the same expression (1.16), (1.17). The probability per unit time for a transition of a spin $I = {}^1/_2$ from the parallel to the anti-parallel state is according to (I.15)

$$W = 2\,\pi\,\gamma^2\,\varrho\,(\nu_0) \qquad (1.17)$$

The resonance phenomenon is obviously just an „optical" magnetic dipole transition. The frequency, however, lies in the broadcast rather than the visible range.

1.5. *The molecular beam magnetic resonance method.*

G o r t e r (G 4) was the first to point out how the phenomenon, described in the previous section, could be used to detect nuclear magnetism. The first successful experiment, however, was performed by R a b i (R 2) with his marvellous molecular beam technique.

Molecules evaporated from the furnace O, pass through some diaphragms to define a beam. The beam is split in the inhomogeneous field of magnet A, passed through a homogeneous field H_0 in magnet C, and is refocused onto a detector by an inhomogeneous field of magnet B, which deflects in the opposite direction as A. The refocusing condition is fulfilled only, if no reorientation of the nuclear spin occurs in magnet C. If a radio frequency magnetic field is applied, perpendicular to H_0, and either the radio frequency or H_0 is slowly changed, the current reaching the detector will pass through a minimum, when the resonance condition (1.7) is fulfilled. Magnetogyric ratio's of many

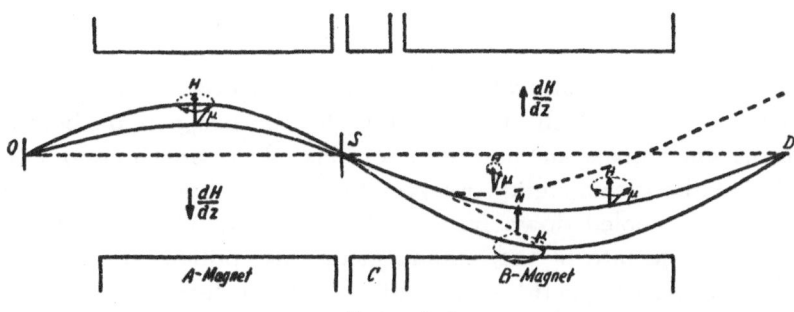

Figure 1.4.

Molecular beam magnetic resonance method, according to Rabi, Millman, Kusch and Zacharias, Phys. Rev. 55, 526, 1939.

Schematic representation of the paths of molecules in which the z-component of the magnetic moment of one of the nuclei has increased, decreased or remained unaltered in the region of magnet C.

nuclei have been measured in this way with great accuracy. The method is usually called the molecular beam method, although it can also be applied to atoms. A special and independent application to neutrons was made by Bloch and Alvarez (A 2). For further information and literature the reader is referred to the review article by Kellogg and Millman (K 2).

1.6. Nuclear Paramagnetism.

The permanent moments of the nuclei should be a source of paramagnetism. The theory of electronic paramagnetism (V 1) can be readily applied to nuclear moments, if one keeps in mind the similarity in the vector diagrams of $\vec{L}$, $\vec{S}$ and $\vec{J}$ and $\vec{J}$, $\vec{I}$ and $\vec{F}$. The contribution of the nuclei to the volume susceptibility is thus given by the well known Langevin formula

$$\chi_0 = N |\mu|^2 / 3 k T \qquad (1.18)$$

where N is the number of nuclei per unit volume, and $|\mu|^2$ denotes the square of the absolute value of the magnetic moment

$$|\mu|^2 = \gamma^2 \hbar^2 I (I + 1) \qquad (1.19)$$

Now it has to be borne in mind that the nuclear moments are about 10^3 times smaller than the electronic ones. Therefore nuclear susceptibilities are roughly a million times smaller than electronic ones. At room temperature the nuclear paramagnetic volume susceptibility of a solid would be of the order 10^{-9} erg / oersted^2, thus negligibly small compared to the ever present diamagnetic volume susceptibility of the order of 10^{-6} erg / oersted[2].

Only at very low temperatures might an influence of nuclear paramagnetism be detected. In the literature has been reported about one experiment which gives an indication for this effect. S h u b n i k o f f (L 1) showed that the diamagnetic susceptibility of solid hydrogen decreased by $20^0/_0$, when cooled down from 4.2° K to 1.7° K. This was attributed by him to an increase of the paramagnetic susceptibility of ortho-hydrogen. The paramagnetic susceptibility arising from the rotational magnetic moment of the molecules is only $3^0/_0$ of the nuclear paramagnetism. The experiment yields a value for the proton moment lying between 2.3 and 2.7 nuclear magnetons. It is to be noted that hydrogen with its few electrons (small diamagnetic effect) and its high density of protons with their relatively large nuclear moment is an exceptionally favorable case. For other substances still lower temperatures would be required. It is almost superfluous to note that saturation effects in nuclear magnetism would occur only at extremely low temperatures. The decisive quantity is $\gamma \hbar H_0/k\,T$, which at room temperature is 7.10^{-6} for protons in a field of 10^4 oersted. Therefore the L a n g e v i n formula (1. 18) should be valid down to about 0.01 $^\circ$K.

1.7. *Nuclear magnetic resonance absorbtion and dispersion.*

G o r t e r (G 5) remarked that, just as e.g. in Na-vapor an anomalous electric dispersion occurs at the position of the yellow resonance line, there must be an anomaly in the nuclear paramagnetic susceptibility in the radio frequency range, if the substance is placed in a large magnetic field H_0. The anomalous dispersion, accompanied by an absorption, will occur at the frequency $\nu_0 = \gamma\,H_0/2\,\pi$.

The susceptibility will, roughly, increase by a factor $\nu_0/\triangle\nu = H_0/\triangle H$, where $\triangle\nu$ is the width of the resonance line. We shall see later that the line width is of the order of a few oersted or less. Therefore the volume susceptibility given by (1, 18) can be increased by a factor 10^4 at resonance or more and would there suddenly jump from practically zero to about 10^{-5}. In 1942 G o r t e r (G 5) attempted in vain to detect this change in susceptibility at low temperatures. The tank coil of an oscillator, filled with KF or LiCl powder, was placed in a field H_0. At the resonance of F^{19} and Li^7 a sudden change in the inductance of the coil should have produced an observable change in the oscillator frequency. Already six years earlier another unsuccesful attempt (G 4, G 6) has been made by the same author to detect the absorption by a resulting rise in temperature of the sample.

The absorption can be described in the following way. In each transition from a level to the next higher one an energy $h\nu = \gamma\hbar H_0$ is absorbed by the nucleus, in the reverse process the nuclear spin system looses the same amount. The transition probability for absorption and stimulated emission is the same. But if the nuclear spin system is in thermal equilibrium there will be more nuclei in the states with lower energy. Thus there will be more transitions up than down, resulting in a net absorption.

We make the table

Level	Energy	Boltzmann factor
m_I	$-\gamma\hbar m_I H_0$	$exp\,(+\gamma\hbar m_I H_0 / kT)$

Since $h\nu = \gamma\hbar H_0 \ll kT$. as we have seen in section 6, there is a constant difference n in population between two adjacent levels, which is small compared to the total number of nuclei N;

$$n = \frac{\gamma\hbar H_0}{kT}\,\frac{N}{2I+1} \qquad (1.20)$$

We have to multiply (1.15) by (1.20) to get the net surplus number of transitions up per second. If we then multiply by $h\nu$, we find for the absorbed power

$$P = {}^4/_3\,\pi\,\gamma^2\,\varrho\,(\nu_0)\,(\gamma\hbar H_0)^2\,N\,I(I+1)/kT \qquad (1.21)$$

The first positive effect of nuclear magnetic resonance absorption was obtained at the end of 1945 by Purcell, Torrey and Pound (P 7), soon afterwards, independently, by Bloch, Hansen and Packard (B 6). The resonance effect in the paramagnetic susceptibility is able to give at least as accurate information on nuclear moments as the molecular beam method. The most important new result for nuclear physics so far obtained with this new method is the magnetic moment and spin of H^3 (A 3, B 5), and the redetermination of the ratio of the moments of proton and neutron and proton and deuteron (A 4). The subject of the present thesis, however, is not to find resonances in a series of isotopes, but to investigate the interaction of the nuclei with one another and with other components of the surrounding substance.

For while in the beam method each particle can be considered as free, in the case of solids, liquids and gases the interaction between the nuclei and their surroundings cannot be neglected. They are, in fact, essential. If it were not for these interactions, the energy absorbed by the nuclei under the influence of a radio frequency magnetic field could not be dissipated and the temperature of the nuclear spin system, entirely isolated from the rest of the sample, would rise. The surplus number in the lower state would decrease, and soon the absorption and stimulated emission would be equal and in the stationary state no net absorption would take place. In the case $I = \frac{1}{2}$ the differential equation for the surplus number n would be

$$\frac{dn}{dt} = -2\,W\,n \qquad (1.22)$$

with the solution

$$n = n\,(o)\,exp\,(-2\,Wt) \qquad (1.23)$$

In this formula W is given by (1.17). The factor two is inserted, because in one transition the surplus number changes by two. With the relation (1.20) we find immediately that the temperature of the system of free spins would increase exponentially

$$T = T_{t\,=\,t_0}\,exp\,(+2\,Wt) \qquad (1.24)$$

To avoid this heating up of the nuclear spin system, a process of energy dissipation must be taken into account. It will be shown in chapter 2 how the interaction between nuclei provides such a process by which thermal equilibrium is restored. The shape and the width of the resonance line will be shown to depend also on these interaction mechanisms, as one would expect. In chapter 3 the experimental method, developed by Purcell's group at Harvard University, will be described. In chapters 4 and 5 the theory will be applied to various substances and compared with experimental results obtained with nuclear resonances of H^1, H^2, Li^7 and F^{19} in these substances. It must be noted that actually a considerable part of the theory was developed after the experiments had been carried out, although sometimes the reverse was true. In order to present a readable account it proved necessary to deviate from the historical development in this thesis.

CHAPTER 2

THEORY OF THE NUCLEAR MAGNETIC RESONANCE.

2. 1. *Rigorous classical solution for free spins.*

The description given in section 1. 4 of the transitions of a free spin in a rotating magnetic field needs some rectification. It is clear that, when in the classical picture the gyroscope has been turned over, the resonating field will start to turn it back, aud the result will be an oscillatory motion. The gyroscope will assume a nutation. The most convenient expression for the equation of motion of this problem is that the rate of change of angular momentum $\vec{J}$ equals the torque exerted by the magnetic field:

$$\frac{d\vec{J}}{dt} = [\vec{\mu} \times \vec{H}]$$

or with (1. 4):

$$\frac{d\vec{\mu}}{dt} = \gamma [\vec{\mu} \times \vec{H}] \qquad (2.1)$$

In a constant field $H_x = H_y = 0$, $H_z = H_0$ the solution of these equations is simply

$$\begin{aligned}
\mu_x &= A \cos (\omega_0 t + \varphi) \\
\mu_y &= A \sin (\omega_0 t + \varphi) \\
\mu_z &= B
\end{aligned} \qquad (2.2)$$

with $\omega_0 = \gamma H_0$, and $A^2 + B^2 = \mu^2$. Classically tg $\vartheta = A/B$ can have any value. From quantummechanics we have the restriction that $\cos \vartheta = m_I / \sqrt{I(I+1)}$. If we have a large number of nuclei which combine to a large total quantumnumber I, the quantummechanical condition will not be severe and the total magnetic moment of the system of nuclei will be correctly described by (2. 1). Torrey showed that an exact solution can also be obtained for the rotating field (1.6) by transforming to a coordinate system that rotates with the same angular velocity $\vec{\omega}$ as the magnetic field. The equations of motions in the new, primed system are (J 2):

$$\frac{D}{dt}\,\vec{\mu'} = \gamma[\mu' \times (\frac{\vec{\omega}}{\gamma} + \vec{H'})] \qquad (2.3)$$

where $\frac{D}{dt}$ denotes differentiation with respect to time in the moving system. In the primed system the magnetic field is constant $H_x' = H_1$, $H_y' = 0$, $H_z' = H_0$.

So $\vec{\mu'}$ will precess about the fixed vector $+ \vec{H'} - \vec{\omega}/\gamma$ according to equation (2.1) with an angular velocity

$$\lambda = \sqrt{(\pm\,\omega + \gamma\,H_0)^2 + \gamma^2 H_1^2} = \sqrt{(\pm\,\omega + \omega_0)^2 + \omega_1^2} \qquad (2.4)$$

In this formula the sign of ω is positive if $\vec{\omega}$ points in the same direction as $\vec{H_0}$. At resonance $\omega = \omega_0$.

The precession becomes a nutation in the original system. We can impose various initial conditions on the general solution of the differential equation. If at $t = 0$ the moment $\vec{\mu}$ is parallel to the vector $\vec{H'} - \vec{\omega}/\gamma$, there is no nutation. In this case we have a pure rotation in the resting system with an angle ϑ given by

$$tg\,\vartheta = \frac{H_1}{-\,\omega/\gamma + H_0} = \frac{\omega_1}{-\,\omega + \omega_0} \qquad (2.5)$$

At resonance we have $\vartheta = \pi/2$. If we start with a constant field $H_z = H_0$ and $\vec{\mu}$ aligned in the z-direction, and at $t = 0$ suddenly apply a rotating component H_1, the moment $\vec{\mu}$ will start to precess with an angular velocity λ and angle $\vartheta = $ arc $tg\ \omega_1/(\omega_0 - \omega)$. Then it follows from simple geometrical considerations, that the z-component of $\vec{\mu}$ as a function of time in the original system is given by

$$\mu_z = \mu \cos^2 \vartheta + \mu \sin^2 \vartheta \cos \lambda t$$
$$\mu_z = \mu (1 - 2 \sin^2 \vartheta \sin^2 {}^1/_2 \lambda t) \qquad (2.6)$$

2. 2. *Rigorous quantummechanical solution for free spins.*

The quantummechanical solution (1. 14) also needs some revision. The application of perturbation theory is only valid for a short period of time, at the end of which the probability of finding the system in the original state is still of the order of unity. In other words (1. 14) holds only as long as $w \ll 1$. An exact solution valid for any time t has been given by Rabi (R 1) in the case $I = \frac{1}{2}$. The problem is to solve the time dependent Schrödinger equation

$$H_{op} \, \psi \, (t) = \frac{\hbar}{i} \frac{\partial \, \psi \, (t)}{\partial \, t} \tag{2.7}$$

where the Hamiltonian H_{op} is given by (1. 9), and the wave function $\psi \, (t)$ has two components

$$\psi \, (t) = c_{1/2} \, (t) \, \psi_{1/2} + c_{-1/2} \, (t) \, \psi_{-1/2} \tag{2.8}$$

The normalised spin wave functions $\psi_{1/2}$ and $\psi_{-1/2}$ belong to the eigenvalues $m_I = \frac{1}{2}$ and $m_I = -\frac{1}{2}$; $| \, c_{1/2} \, (t) \, |^2$ and $| \, c_{-1/2} (t) \, |^2$ are the probabilities to find the system at time t in the state $m_I = +\frac{1}{2}$ or $m_I = -\frac{1}{2}$ respectively. Since the system must be in one state or another, we have the normalisation condition

$$| \, c_{1/2} \, (t) \, |^2 + | \, c_{-1/2} \, (t) \, |^2 = 1 \tag{2.9}$$

Equation (2. 7) can be written in matrix form

$$\frac{\hbar}{i} \begin{pmatrix} \dot{c}_{1/2} \\ \dot{c}_{-1/2} \end{pmatrix} = -\gamma \, \hbar \begin{pmatrix} H_0 & H_x + i H_y \\ H_x - i H_y & H_0 \end{pmatrix} \begin{pmatrix} c_{1/2} \\ c_{-1/2} \end{pmatrix} \tag{2.10}$$

These two simultaneous differential equations can be solved with the initial conditon $| \, c_{1/2} \, (t) \, |^2 = 1$ at $t = 0$.

Rabi finds for the probability that the system is in the state $m_I = -\frac{1}{2}$ at time t

$$| \, c_{-1/2} \, (t) \, |^2 = w_{-1/2, \, 1/2} = \frac{\omega_1{}^2}{(\omega - \omega_0)^2 + \omega_1{}^2} \sin^2 \frac{t}{2} \sqrt{(\omega - \omega_0)^2 + \omega_1{}^2} \tag{2.11}$$

This correspondends exactly to the classical solution (2. 6), if one remembers that for spin $\frac{1}{2}$, $\mu_z = \mu \, (1 - 2 \, w_{-1/2, \, 1/2})$.

Schwinger (S 3) showed that for obtaining the quantummechanical solution it also has some advantage to transform to a rotating coordinate system. The results can be extended to the case of arbitrary spin by the general M a j o r a n a formula, which one can find in R a b i's paper (R 1, B 8). We note that (2. 11) goes over into (1. 14) for very small values of the perturbation field $\omega_1 \ll \omega - \omega_0$. At resonance $\omega = \omega_0$, $w_{-1/2, 1/2}$ becomes equal to unity for $t = \pi/\omega_1$. The system then oscillates between the states with spin parallel and antiparallel to the magnetic field. If $\omega \neq \omega_0$, the system never attains the pure state with $m = -1/2$, and oscillates more rapidly than at resonance for the same value of H_1. We can define the width of the resonance curve as the distance between the points, where the maximum chance to find the system in the antiparallel state is one half. From (2. 11) we see that $|\omega - \omega_0| = \omega_1$ for those points, or the width measured in oersted is twice the amplitude of the radiofrequency field. The energy of the spin system in the magnetic field H_0 (at $t = 0$, all n spins are parallel to the field) is given by

$$E(t) = -1/2 (1 - 2w) \gamma \hbar H_0 n \qquad (2. 12)$$

and the time derivative of this expression yields the absorbed power, which behaves as $\sin t \sqrt{(\omega - \omega_0)^2 + \omega_1^2}$. In the average no net absorption of energy takes place. Suppose now that we have an assembly of free spins in a not perfectly homogeneous magnetic field. The distribution function of the spins over the resonance frequencies be $\varphi(\nu)$ which is normalised $\int_0^\infty \varphi(\nu) d\nu = 1$. We assume that the distribution over the inhomogeneous field is much wider than the width of the resonance, which, expressed in oersted, is about H_1. We assume that $\varphi(\nu)$ is practically constant near the resonance $\varphi(\nu) \approx \varphi(\nu_0)$ for $|\nu - \nu_0| < \nu_1$. Then the energy absorbed by the system is

$$E(t) = \int_0^\infty w_{-1/2, 1/2}(\nu) \gamma \hbar H_0 n \varphi(\nu) d\nu =$$

$$= \frac{1}{2\pi} \gamma \hbar H_0 \varphi(\nu_0) n \int_0^\infty \frac{\omega_1^2 \sin^2 \frac{t}{2} \sqrt{(\omega - \omega_0^1)^2 + \omega_1^2}}{(\omega - \omega_0^1)^2 + \omega_1^2} d\omega_0^1 \quad (2. 13)$$

Take the time derivative, representing the power absorbed, of this expression. The integral can then be evaluated in terms of the Bessel function of zero order.

$$\frac{dE(t)}{dt} = P(t) = 1/4 \omega_1^2 \gamma \hbar H_0 \varphi(\nu_0) n J_0(\gamma H_1 t) \qquad (2. 15)$$

For small t we find, substituting $J_0(0) = 1$, in (2. 15)

$$P(t) = \frac{1}{4}\, \gamma^2\, H_1^2\, \varphi\,(\nu_0)\, n\, \gamma\, \hbar\, H_0 \qquad (2.16)$$

It is not surprising that this comes out to be the same as if we had used a time proportional transition probability given by (1. 17) to calculate the absorbed power.

In the derivation of that formula we had supposed a range of frequencies in the radio frequency signal, over which we had to integrate, as is usually done in optics. In the case of a single applied frequency, but a distribution over H_0, we have to integrate rather over a range of resonance frequencies. Since ω and ω_0 occur only in the combination $\omega - \omega_0$, the result is the same. In r.f. spectra this last case is more common, quite contrary to the situation usually encountered in optics. From (2. 15) we see that in the stationary state here is again no net absorption of energy as the Bessel function vanishes as $x^{-1/2}$ for large arguments. This means that the free spins, initially all oriented in one direction, start oscillating under the influence of the applied signal, get out of phase and for large t there are always as many pointing up as down.

2. 3. *Interaction with radiation.*

In order to keep an absorption in the stationary state, it is necessary that there is some mechanism trying to restore thermal equilibrium so that the population of the spin levels does not become equal. The time T_1, it takes for the spin system to come back to thermal equilibrium, after this has been disturbed some way or other, e.g. by a large radio frequency signal at resonance, is called the relaxation time. The experiments described in chapter 3 and 4 yield values for the relaxation time, ranging from 10^{-4} to 10^2 seconds. We shall now discuss some interaction mechanisms which tend to restore thermal equilibrium. Even for the so called „free" particles one always has the interaction with radiation. The spontaneous emission which usually limits the life time for an atom in a electronically excited state to 10^{-8} sec, is negligibly small for radio frequency transitions. The coefficient for spontaneous emission A of dipole radiation (H 1) is proportional to the cube of the frequency and the square of the dipole moment.

$$A = 8\,\pi\,h\,\nu^3\,B/c^3 \qquad (2.17)$$

where B is the coefficient of absorption or induced emission, i.e. w in (1. 15), is $\varrho\,(\nu_0)$ and t are taken to be unity. Substituting numerical values for protons in a field of 10^4 oersted one finds $A = 10^{-25}$ sec^{-1}, corresponding to a life time of 10^{19} years. This is not the relaxation time caused by the radiative interaction. For, in addition to the spontaneous emission, we have the transitions induced by the thermal radiation field. Take $I = \frac{1}{2}$, and let n^+ and n^- denote the number of spins in the upper and lower level respectively, n^+ satisfies the differential equation

$$\frac{dn^+}{dt} = B \varrho(\nu) n^- - \{A + B\varrho(\nu)\} n^+ \qquad (2.18)$$

Using $n^+ + n^- = N$ we find

$$n^+ = C e^{-\{A + 2 B \varrho(\nu)\} t} + \frac{B \varrho(\nu)}{A + 2 B \varrho(\nu)} N \qquad (2.19)$$

Thus the relaxation time T_1, if radiation were the only interaction, would be $\{A + 2 B \varrho(\nu)\}^{-1}$. Here $\varrho(\nu)$ is the energy density of the electromagnetic field. As for the nuclear resonance $h\nu/kT \ll 1$, we may use Rayleigh's approximation

$$\varrho(\nu) = 8 \pi \nu^2 k T c^{-3} \qquad (2.20)$$

With $B = 2 \pi \gamma^2$, we find $T_1 \approx 10^{18}$ years. Although the influence of the thermal radiation field is much larger than that of the spontaneous emission at these low frequencies, the effect of the radiation is far too small to play any role in the explanation of the observed relaxation times. Purcell (P 4) pointed out that the energy density of blackbody radiation is not given by Rayleigh's formula, if the wave length is large compared to the dimensions of the black body. This is exactly the case at radio frequences in a tuned LC circuit. The energy density at the resonance frequency is increased, because the energy kT of the circuit is stored in a narrow frequency range.

The mean square voltage across a resistor of temperature T is $\overline{V^2} = 4 R k T \triangle \nu$.

At resonance the current is

$$\overline{i^2} = 4 k T \triangle \nu / R$$

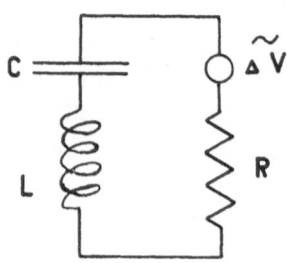

Figure 2. 1.
Noise in a tuned circuit.

The energy stored in the coil is $\frac{1}{2} L \overline{i^2}$. Introduce $Q = \omega L/R$ and the volume of the coil V. We find for the energy density of the supposedly homogeneous magnetic field per unit frequency range

$$\varrho(\nu) = k T Q/\pi \nu V \qquad (2.21)$$

The increase over the energy density in a large black-body cavity is given by the factor $\lambda^3 Q/8 \pi^2 V$. Substituting $V = 1$ cc, $\lambda = 10^3$ cm, $Q = 100$, we find that the relaxation time is decreased by a factor 10^9, but is still 10^8 years, so several orders of magnitude above the experimental values.

2. 4. Dipole — dipole interaction.

Looking for other types of interaction besides the very small radiation damping, the contact of the nuclear spin system with the outside world appears to be very limited. Electric forces, which act during atomic and electronic collisions and play e.g. an important role in the mechanism of arcs, readily establish an equilibrium for electronic states. But they do not perturb the nuclear spin. Only a gradient of an electric field can interact with a nuclear quadrupole moment, as will be considered in chapter 5.

We shall neglect the possibility of interaction by exchange forces. For electronic states with overlapping wave functions this exchange is sometimes considerable (ferromagnetism or antiferromagnetism). But for the nuclei in crystals and liquids at room temperature exchange is not likely to occur.

So we come back to a magnetic type of interaction, by means of the magnetic moment associated with the nuclear spin. This interaction will of course, be much smaller than the corresponding magnetic interaction of electronic moments, as the nuclear magneton is so much smaller than the Bohr magneton. So far the magnetic field acting on a nucleus was taken to consist only of the externally applied field $H_0 + H_1$ and the thermal radiation field. Every nucleus, however, also experiences the field produced by the magnetic moments of its neighbours. If this dipole — dipole interaction is taken into account, the Hamiltonian (1. 9) for an assembly of spins becomes

$$H_{op} = \Sigma_i \, \gamma_i \, \hbar \vec{I}_i \cdot \vec{H}_0 + \Sigma_i \, \gamma_i \, \hbar \vec{I}_i \cdot \Sigma_j \left(\frac{\gamma_j \hbar \vec{I}_j}{r^3_{ij}} - \frac{3 \gamma_j \hbar \, \vec{r}_{ij} (\vec{r}_{ij} \cdot \vec{I}_j)}{r^5_{ij}} \right) \quad (2.24)$$

The sum over j in the right hand term represents the internal or local field H_{loc} at the ith nucleus; $\vec{r}_{ij} = \vec{r}_j - \vec{r}_i$ is the radiusvector connecting the ith and the jth spin. The problem connected with (2.24) is one of an extreme complexity. It is the equation of motion for N particles, where N is of the order of the number of Avogadro. The local field will, at any time, be of the order of few oersted. It is mainly determined by the nearest neighbours, since the magnitude of a dipole field decreases as the inverse cube of the distance: $H_{loc} \approx \gamma \hbar / a^3$ where a is the internuclear distance. For two protons, one Ångstrøm apart, we find $H_{loc} \approx 10$ oersted, The problem (2.24) of an assembly of spins also comes up in the theory of the absorption and dispersion of electronic magnetism in paramagnetic salts. G o r t e r (G 3) gives a survey of experiment and theory in this field. The reader will find in this book ample references to the existing literature. Using the classification customary in electronic magnetism, the case which is of interest here is that of no electric splitting and a strong transverse field. It must be stressed that we are interested in a resonance absorption. Recently the first experiments of electronic magnetic resonance at microwave frequencies have been reported (C 6). Most of the work on absorption and dispersion in paramagnetic crystals, however, relates to the non-resonant absorption. We shall here proceed on similar lines as B r o e r (B 12).

The energy levels of the unperturbed system, without dipole-dipole interaction are (compare (2.12))

$$E_n = \tfrac{1}{2} \, n \, \gamma \, \hbar \, H_0 \qquad\qquad (2.25)$$

If $I = \tfrac{1}{2}$, $n = N^+ - N^-$ is the difference between the number of spins which are parallel or antiparallel to H_0. Since n is even or odd, depending on whether the total number $N = N^+ + N^-$ is even or odd, the spacing between the equidistant levels is $\gamma \hbar H_0$. The degeneracy of the levels is high, $N!/N^+! N^-!$, but is lifted by the interaction term in (2.24). Now the perturbation energy will be of the order $N^{1/2} \gamma \hbar H_{loc}$. Although $H_{loc} << H_0$, the perturbation energy for an assembly of many

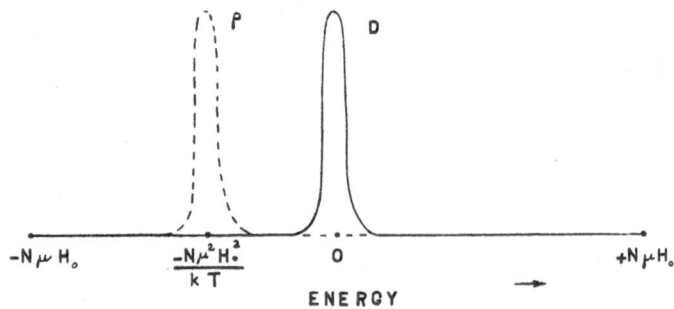

Figure 2.2.

The distribution D of energy levels of a spin system in a magnetic field H_0 and the distribution of occupied states ϱ, obtained by multiplying D with the Bolzmann factor $e^{-Hop/kT}$. Both functions are normalised.

spins ($N \approx 10^{22}$) will be large compared to the spacing of the unperturbed levels. In view of the large number of non-degenerate levels we can say that the energy levels belonging to (2.24) are spread over a continuum. The density D of the energy levels as a function of the energy has sharp maximum, the width being of the order $N^{1/2} \gamma \hbar H_0$. The average occupation ϱ of the continuum is obtained by multiplying D by the Bolzmann factor exp $(- H_{op}/k T)$. The regular spacing $\gamma \hbar H_0$ has completely vanished from the picture. It seems as if the whole resonance phenomenon disappeared, for one might expect transitions from any level p to any level q in the continuous distribution, when a rotating magnetic field H_1 is applied with frequency $\nu = E_p - E_q /h$. The probability for such a process per unit time is

$$
\begin{aligned}
W_{q \leftarrow p} &= {}^1\!/_4 \gamma^2 H_1^2 \mid (p \mid \Sigma_j (I_x + i I_y)_j \mid q) \mid^2 \varrho_q \\
&= \frac{1}{4 \hbar^2} H_1^2 \mid M_{pq} \mid^2 \varrho_q
\end{aligned}
\tag{2.26}
$$

where M_{pq} is the matrix element of the rotating component of the total magnetic moment between the states p and q. The absorbed power $P(\nu)$ is obtained by subtracting from (2.26) the transitions from q to p, multiplying by the involved energy $h \nu$ and summing over a small frequency interval $2 \triangle \nu$ around ν. For $h \nu/k T \ll 1$ we can write

$$
\varrho_p - \varrho_q = \varrho_{pq} \, h \nu/k T
\tag{2.27}
$$

with $\varrho_{pq} \approx (\varrho_p + \varrho_q)/2$

$$
P(\nu) = \frac{H_1^2}{4 \hbar^2} \frac{h^2 \nu^2}{k T} \sum_{\nu - \triangle \nu}^{\nu + \triangle \nu} \mid M_{pq} \mid^2 \varrho_{pq}
$$

Introducing the absorption coefficient $A(\nu) = \dfrac{P(\nu) \, 16 \pi}{H_1^2}$ and the density function of the magnetic moment

$$
f(\nu) = 2 \sum_{\nu - \triangle \nu}^{\nu + \triangle \nu} \mid M_{pq} \mid^2 \varrho_{pq}
\tag{2.28}
$$

we find

$$
A(\nu) = \frac{8 \pi^3 \nu^2}{k T} f(\nu)
\tag{2.29}
$$

These last two equations are identical with those of B r o e r (Thesis,

p. 63). The problem is reduced to determining $f(\nu)$. The only exact way for solving this problem is the diagonal sum method, developed by Waller, Van Vleck and Broer (W 1, V 2, B 13). Neglecting terms in $h\nu/kT$, they find the relations

$$\int_0^\infty f(\nu)\,d\nu = \text{Spur } M^2$$

$$\int_0^\infty \nu^2 f(\nu)\,d\nu = \text{Spur } \dot{M}^2/4\,\pi^2$$

$$\int_0^\infty \nu^4 f(\nu)\,d\nu = \text{Spur } \ddot{M}^2/16\,\pi^4 \quad \text{etc.} \tag{2.30}$$

Furthermore we have the general relation for the time derivative of a quantummechanical operator

$$i\,\hbar\,\dot{M} = HM - MH \tag{2.31}$$

where H is the Hamiltonian.

The mean square frequency of the absorption curve $A(\nu)$ is the quotient of the first two expressions (2.30)

$$\overline{\nu^2} = \frac{\text{Spur}(HM - MH)^2}{h^2\,\text{Spur } M^2} \tag{2.32}$$

The shape of the curve remains obscure. Strictly speaking, one does not even know whether there is a resonance at all. Unfortunately the the quantities $\overline{\nu^4}$, $\overline{\nu^6}$ etc. are hard to calculate, although in principle they will give more information about the shape of the absorption curve (V 3).

Our knowledge is supplemented by the perturbation method. At first it may seem strange that a perturbation theory can be applied, since we have seen that the perturbation energy is larger than the spacing of the unperturbed levels. However, as Broer pointed out, the matrix elements of the magnetic moment operator differ appreciably from zero only when the energy difference $h\,\triangle\nu$ between two states p and q satisfies the relation $h\,\triangle\nu \approx \gamma\,\hbar\,H_{loc}$ or $h\,\triangle\nu \approx \gamma\,\hbar(H_0 \pm H_{loc})$ or $h\,\triangle\nu \approx \gamma\,\hbar\,(2\,H_0 \pm H_{loc})$.

The schematic behaviour of $f(\nu)$ is plotted in fig. 2.3. The absorption can be obtained from (2.29). The absorption at the Larmor frequency is a first order effect, in which we are interested. The absorption near a frequency of zero or 2ν is of the order $(H_{loc}/H_0)^2$ and so about 10^6 times smaller than at ν_0, and unobservable in the case of nuclear magnetism.

The use of the perturbation theory is justified by its results, which

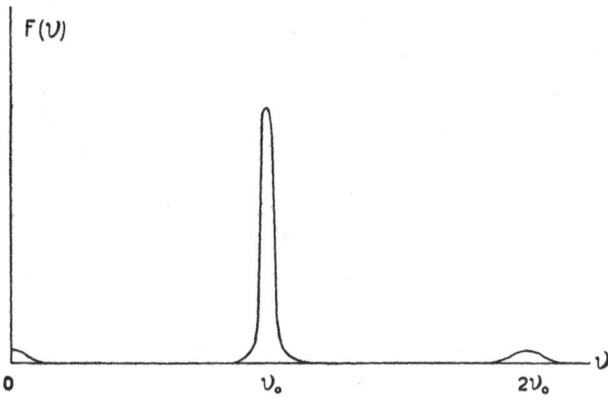

Figure 2. 3.

The density function $f(v)$ of the magnetic moment (cf. Broer, B 12).

are in agreement with the experimental facts, one of which is the occurrence of a sharp resonance at the frequency v_0. The perturbation method gives more details than the diagonal sum method, but is less rigorous and certain. Sometimes it is advantageous to combine them.

The unperturbed state of the system is specified by the quantumnumbers m_{Ij} of all spins, the total z-component of the spins being $m = \sum_j m_j$. We assume that these magnetic quantum numbers still characterize a state after the introduction of the perturbation

$$V = \sum_{j > i} \sum V_{ij}$$

$$V_{ij} = \frac{\gamma^2 \hbar^2}{r^3_{ij}} [I_{x_i} I_{x_j} (1 - 3 a_1^2) + I_{y_i} I_{y_j} (1 - 3 a_2^2) + I_{z_i} I_{z_j} (1 - 3 a_3^2) -$$
$$- 3 (I_{x_i} I_{y_j} + I_{x_j} I_{y_i}) a_1 a_2 - 3 (I_{x_i} I_{z_j} + I_{x_j} I_{z_i}) a_1 a_3 -$$
$$- 3 (I_{y_i} I_{z_j} + I_{y_j} I_{z_i}) a_2 a_3]$$

$(2. 33)$

V_{ij} is the magnetic interaction between the ith and j^{th} spin; a_1, a_2 and a_3 are the direction cosines of the radius vector $\vec{r}_{ij}$ with respect to a system, of which the z-axis is parallel to H_0. We shall now write the operator (2. 33) in the m-representation, so that we can distinguish terms which leave the total z-component m unchanged, $\triangle m = 0$, and terms for which $\triangle m = +1$ or -1 and those for which $\triangle m = +2$ or -2. The elements with $\triangle m = 0$ leave the energy unchanged. With off-diagonal elements is a change in energy $\triangle m \gamma \hbar H_0$ connected. We make use of the relations

$$I_{x_i} I_{x_j} + I_{y_i} I_{y_j} = \frac{(I_{x_i} - i I_{y_i})(I_{x_j} + i I_{y_j})}{2} + \frac{(I_{x_i} + i I_{y_i})(I_{x_j} - i I_{y_j})}{2}$$

$$I_{x_i} I_{x_j} - I_{y_i} I_{y_j} = \frac{(I_{x_i} + i I_{y_i})(I_{x_j} + i I_{y_j})}{2} + \frac{(I_{x_i} - i I_{y_i})(I_{x_j} - i I_{y_j})}{2}$$

$$I_{x_i} I_{y_j} + I_{x_j} I_{y_i} = \frac{(I_{x_i} + i I_{y_i})(I_{x_j} + i I_{y_j})}{2 i} - \frac{(I_{x_i} - i I_{y_i})(I_{x_j} - i I_{y_j})}{2 i}$$

$$I_{x_i} I_{z_j} \qquad = {}^1\!/_2 \, (I_{x_i} + i I_{y_i}) I_{z_j} + {}^1\!/_2 \, (I_{x_{i,}}' - i I_{y_i}) I_{z_j}$$

$$I_{y_i} I_{z_j} \qquad = \frac{1}{2 i}(I_{x_i} + i I_{y_i}) I_{z_j} - \frac{1}{2 i}(I_{x_i} - i I_{y_i}) I_{z_j}$$

and introduce the polar and azimuthal angles ϑ_{ij}, and φ_{ij}, instead of the direction cosines. We can then write (2. 33) in the form, adding the Heisenberg time factors $\exp(i \triangle m \gamma H t)$,

$$V_{ij} = \frac{\gamma^2 \hbar^2}{r^3_{ij}} \, (A + B + C + D + E + F) \qquad (2.34)$$

$\triangle m = 0$
$$A = I_{z_i} I_{z_j} (1 - 3 \cos^2 \vartheta_{ij})$$

$\triangle m = 0$
$$B = -{}^1\!/_4 \{ (I_{x_i} - i I_{y_i})(I_{x_j} + i I_{y_j}) + $$
$$+ (I_{x_i} + i I_{y_i})(I_{x_j} - i I_{y_j}) \} (1 - 3 \cos^2 \vartheta_{ij})$$

$\triangle m = 1$
$$C = -{}^3\!/_2 \{ (I_{x_i} + i I_{y_i}) I_{z_j} + $$
$$+ (I_{x_j} + i I_{y_j}) I_{z_i} \} \sin \vartheta_{ij} \cos \vartheta_{ij} e^{-i \varphi_{ij}} e^{i \gamma H_0 t}$$

$\triangle m = -1$
$$D = -{}^3\!/_2 \{ (I_{x_i} - i I_{y_i}) I_{z_j} + $$
$$+ (I_{x_j} - i I_{y_j}) I_{z_i} \} \sin \vartheta_{ij} \cos \vartheta_{ij} e^{+i \varphi_{ij}} e^{-i \gamma H_0 t}$$

$\triangle m = 2$
$$E = {}^3\!/_4 \, (I_{x_i} + i I_{y_i})(I_{x_j} + i I_{y_j}) \sin {}^2\!\vartheta_{ij} \, e^{-2 i \varphi_{ij}} e^{2 i \gamma H_0 t}$$

$\triangle m = -2$
$$F = {}^3\!/_4 \, (I_{x_i} - i I_{y_i})(I_{x_j} - i I_{y_j}) \sin^2 \vartheta_{ij} \, e^{+2 i \varphi_{ij}} e^{-2 i \gamma H_0 t}$$

The first two terms give rise to secular perturbations, the last four are periodical perturbations of very small amplitudes. K r a m e r s gives a clear exposition of these two different types of perturbation in his book on quantum mechanics (K 6). That the terms C to F cannot give rise to appreciable changes also follows from the principle of conservation of energy. So we keep only the terms A and B. Classically A corresponds to the change in the z-component of the magnetic field by the z-component of the neighbouring nuclear moments. This local field will vary from nucleus to nucleus, slightly changing the Larmor frequency of each. Let us consider in somewhat more detail the situation at the ith nucleus. Let us suppose for simplicity that the neighbouring nuclei have spin $I = 1/2$. Each of them can be parallel or antiparallel to H_0. For each configuration we have a definite value of the z-component of the local field. Let us first consider only the nearest neighbours. Instead of one value H_0, we have e.g. ten possible values around H_0, depending which neighbours are parallel and which antiparallel to H_0. Now all the configurations of the next nearest neighbours will split each of the ten values into many more, and so on for the third nearest, etc. The result will be a continuum of values of the z-component of the local field and therefore a distribution over a range of Larmor frequencies. In many cases this distribution, approximately will have a Gaussian shape, as was pointed out by several authors (K 10, V 2). But we want to stress that this is not necessarily so. Take for instance the case of a substance in which the nuclei occur in pairs. That is, each nucleus has one very close neighbour, while the next nearest are relatively far away. The nearest neighbour will split the H_0 value into two discrete values, rather far apart. The next nearest neighbours will produce a bell shaped distribution of the Larmor frequency around each of these two discrete values. The nuclear resonance line will show a fine structure in this case. This has been observed experimentally in many crystals by P a k e (P 1). It will be very difficult to obtain information about the fine structure with the diagonal sum method.

So we see that for each crystal a detailed investigation would be required. But as in many cases the line will be a Gaussian anyway, we can obtain an estimate of the line width by calculating the mean square contribution of each nucleus separately, and taking the square root of the sum of these contributions, as the orientation of each nucleus is independent of the others. The mean square local field from term A is given by

$$\overline{(\triangle H_{loc})^2} = \gamma^2 \hbar^2 \frac{I(I+1)}{3} \sum_j \frac{(1 - 3\cos^2 \vartheta_{ij})^2}{r^6_{ij}} \tag{2.35}$$

and the mean square deviation in frequency by

$$\overline{(\triangle\,\omega)^2} = \gamma^2\,\overline{(\triangle\,H_{loc})^2}.$$

To this we have to add the contribution of term B. To B corresponds the simultaneous flopping of two antiparallel spins ($\triangle\,m_i = +1$ and $\triangle\,m_j = -1$ or vice versa). This process is energetically possible and is caused by the precession of the nuclei around H_0 with the Larmor frequency. So they produce an oscillating field of resonance frequency at the position of their neighbours, resulting in reciprocal transitions. Classically one might say that this process limits the life time of the spin in a given state and therefore broadens the spectrum. To (2. 35) we have to add a numerical factor to take the effect of B into account.

The proper factor has been calculated by Van Vleck (V 3) in a rigorous manner with the diagonal sum method, using formula (2. 32). For H_{op} we take the terms of type A and B of the perturbation $\Sigma\,\Sigma\,V_{ij}$. The component of the magnetic moment, rotating with the $j>i$ applied radio frequency field, is proportional to $\Sigma\,I_{x_j} + i\,I_{y_j}$. We can j evaluate the commutator by making use of the commutation rules which exist for the components of the angular momentum operator I. Van Vleck's (V 3) result is

$$(1/T_2')_{asymptotic} \equiv \sqrt{\overline{(\triangle\,\omega)^2}} = {}^3/_2\,\gamma^2\,\hbar\,\sqrt{\frac{I(I+1)}{3}}\,\sqrt{\Sigma_j\frac{(1-3\cos^2\vartheta_{ij})^2}{r^6{}_{ij}}}\,(2.\,36)$$

This formula, valid for a system of identical spins, holds for any shape of the nuclear resonance line. We have no information what the contribution to $\sqrt{(\triangle\,\omega)^2}$ from the tails is. For a Gaussian $(2/T_2')_{asymptotic}$ is the width between the points of maximum slope. The reason for this notation we shall see later.

2. 5. *The relaxation time.*

So far we still have not found a relaxation process. For we argued that only processes with $\triangle\,m = 0$ can occur. To restore thermal equilibrium it is essential that the energy of the spin system changes. We have tacitly assumed that the position coordinates r_{ij}, ϑ_{ij} and φ_{ij} are constants, and we have seen that the dipole-dipole interaction of nuclei at fixed positions only gives a broadening of the levels. Now we shall show that energy transfer is produced, if *the position vectors* $\vec{r}_{ij}$ *are functions of the time* (B 11). This is the only new feature in the treatment of the spin system presented here. In practice the nuclei will always move to some extent, if the system has not zero temperature. In a crystal we have the lattice vibrations, while in a liquid or gas we have the Brownian movement. As far as the Hamiltonian (2.24) for the nuclei is concerned, we can consider this motion of the molecules as to be produced by external forces, since the atomic interactions are mainly of electric nature. In doing so we neglect, of course, the reaction of the magnetic moment of the nuclei on the Brownian motion. We expand the factors in (2.34), which contain the position coordinates and are thus functions of the time, into a series of Fourier or better a Fourier integral. We want to distingiush in our complex notation between positive and negative frequencies and we therefore define the intensity $J(\nu)$ of the Fourier spectra of the functions of the position coordinates

$$F_0 = \sum_{ij} (1 - 3 \cos^2 \vartheta_{ij}) / r^3{}_{ij}$$

$$F_1 = \sum_{ij} \sin \vartheta_{ij} \cos \vartheta_{ij} \; e^{\,i\,\varphi_{ij}} / r^3{}_{ij}$$

$$F_2 = \sum_{ij} \sin^2 \vartheta_{ij} \; e^{\,2i\,\varphi_i} / r^3{}_{ij}$$

by the equations *):

$$\overline{\sum_j \left| \left(1 - 3 \cos^2 \vartheta_{ij}(t)\right) \middle/ r^3{}_{ij}(t) \right|^2} = \int_{-\infty}^{+\infty} J_0(\nu) \, d(\nu) \qquad (2.37)$$

$$\overline{\sum_j \left| \sin \vartheta_{ij}(t) \cos \vartheta_{ij}(t) \, e^{\,i\,\varphi_{ij}(t)} \middle/ r^3{}_{ij}(t) \right|^2} = \int_{-\infty}^{+\infty} J_1 \, \nu \, d(\nu) \quad (2.38)$$

$$\overline{\sum_j \left| \sin^2 \vartheta_{ij}(t) \, e^{\,2\,i\,\varphi_{ij}(t)} \middle/ r^3{}_{ij}(t) \right|^2} = \int_{-\infty}^{+\infty} J_2(\nu) \, d\nu \qquad (2.39)$$

We shall first reconsider the terms C, D, E and F in the expression (2.34) for the perturbation. If $J_2(2\nu_0)$ and $J_2(-2\nu_0)$ are different from zero, E and F become secular perturbation, because the time factor cancels out. Similary C and D become secular perturbations, if $J_1(\nu_0)$

---

*) In section 4.1 the reader will find a short discussion of the Fourier series and spectral intensity of a random function.

and $J_1(-\nu_0)$ do not vanish. The action of the terms C and D may classically be described as transitions induced by the Larmor frequency of the spectrum of the local field. The thermal motion provides the energy necessary for the change $\triangle m$. The interaction of the nuclei with the thermal motion is the relaxation mechanism. From E and F we see, that, quantum-mechanically the simultaneous transition of two nuclei is also a possible process. Let us ask for the probability for a change $\triangle m_i = +1$ in the magnetic quantumnumber of the ith spin. We repeat the simple perturbation calculation of chapter 1 with the terms C and E rather than (1.12): Taking the average over all values m_j of the neighbours we find for the transition probability,

$$W_{m_i+1 \leftarrow m_i} = {}^9/_{16}\,\gamma^4\,\hbar^2\,J_2\,(-2\,\nu_0)\,{}^2/_3\,I\,(I+1)\,(I-m_i)\,(I+m_i+1) +$$
$$+ {}^9/_4\,\gamma^4\,\hbar^2\,J_1\,(-\nu_0)\,{}^1/_3\,I\,(I+1)\,(I-m_i)\,(I+m_i+1) \quad (2.40)$$

It is simple to generalize (2.40) to the case that the spins and magnetogyric ratio's of the nuclei are not all the same

$$W_{m_i+1 \leftarrow m_i} = {}^3/_8\,\gamma_i^2\hbar^2\,(I-m_i)(I_i+m_i+1)\,\sum_j\,[J_{2j}\,(-\nu_{0_i}-\nu_{0_j}) +$$
$$+ 2\,J_{1j}\,(-\nu_{0_i})] \quad (2.41)$$

where J_{2j} stands for the intensity of

$$\gamma_j\,\sqrt{I_j(I_j+1)}\,\sin^2\vartheta_{ij}\,e^{2\,i\varphi_{ij}}/r^3_{ij}.$$

The neighbouring moments may now also be caused by molecular rotation or they may consist of electronic spins and orbital momenta. In these cases not only the position coordinates may be a function of the time, but also the quantisation of the spin $m_j(t)$ must be regarded as time dependent under the influence of external forces. Instead of (2, 41) we should write

$$W_{m_i+1 \leftarrow m} = \gamma_i^2\,\hbar^2(I_i-m_i)(I_i+m_i+1)\,[J''\,(-\nu_{0_j}-\nu_{0_i}) + J'\,(-\nu_{0_i})] \quad (2.42)$$

where $J''(\nu)$ is the intensity of the function

$$\sum_j\left[\sqrt{{}^3/_8\,[I_j-m_j(t)][I_j+m_j(t)+1]}\,\sin^2\vartheta_{ij}(t)\,e^{2\,i\varphi_{ij}(t)}/r^3_{ij}(t)\right]$$

and $J'(\nu)$ is the intensity of

$$\sum_j {}^1\!/_2 \sqrt{3}\, m_j\,(t) \sin \vartheta_{ij}\,(t) \cos \vartheta_{ij}\,(t)\, e^{i\,\varphi_{ij}\,(t)}\, /r^3{}_{ij}\,(t)$$

These intensities represent the spectrum of the local magnetic field.

Identical expressions exist for transitions with $\triangle m_i = -1$. In that case the intensities have to be taken at positive frequencies in the Fourier spectrum. We shall see later that $J(\nu)$ is an even function. This is plausible, since the negative frequencies have no specific physical meaning and are merely a consequence of complex notation. One might object that on this basis the thermal motion would produce as many transitions upward as downward, and still the desired energy dissipation would not occur. Since we have to do, however, with a thermal mechanism, it is appropriate to weight the transition probabilities with the Bolzmann factor of the final state. Dealing with the interaction with radiation, Einstein did essentially the same thing by introducing the spontaneous emission, so as to increase the number of downward transitions. This procedure can be justified by postulating that in the case of equilibrium we have a Bolzmann distribution over the energy levels and a detailed balance for each transition process (G 6).

$$N_q\, W_{p \leftarrow q} = N_p\, W_{q \leftarrow p}$$

$$N_q/N_p = \exp\,(E_p - E_q)/k\,T \qquad (2.43)$$

$$W_{p \leftarrow q}/W_{q \leftarrow p} = \exp\,(E_q - E_p)/k\,T$$

In the case that $I = {}^1\!/_2$ we can write down the differential equation for the population in the upper and lower state in the same way as we did in dealing with the interaction with radiation (section 2.3)

$$\frac{d\,N^+}{d\,T} = -W\,N^+ exp\,\frac{\gamma\,\hbar\,H_0}{2\,k\,T} + W\,N^-\,exp-\frac{\gamma\,\hbar\,H_0}{2\,k\,T}$$

$$N^+ + N^- = N \qquad (2.44)$$

$$W \approx (W_{\triangle m = +1} + W_{\triangle m = -1})/2$$

Expanding the exponentials and keeping only terms of the first order in $\gamma\,\hbar\,H_0/k\,T$, the solution becomes

$$N^+ = C\,e^{-2\,W\,t} + {}^1\!/_2\,N\,(1 - \gamma\,\hbar\,H_0/2\,k\,T) \qquad (2.45)$$

The constant C is determined by the initial conditions at $t = 0$. So

N^+ approaches its equilibrium value asymptotically according to an exponential function with a characteristic time given by

$$T_1 = 1/2 \ W \qquad (2.46)$$

The surplus number $n = N - 2 N^+$ has the same characteristic relaxation time. For W we shall take the value given by the formulæ (2.40) to (2.42), although this is not quite correct for the processes with $\triangle m = \pm 2$, nor for the case, that there are different magnetogyric ratios.

Then we should solve a more complicated system of simultaneous differential equations.

The case that $I > {}^1/_2$ is more complicated. If there are only two possible orientations of the spin, a temperature of the spin system can always be defined by the relation $N^+/N^- = \exp(-\gamma \hbar H_0/k \ T)$.

If $I > {}^1/_2$, a temperature can only be defined if

$$N_I/N_{I-1} = N'_{I-1}/N_{I-2} = \ldots \qquad = \exp(-\gamma \hbar H_0/k \ T) \qquad (2.47)$$

We require that this condition is fulfilled at $t = 0$. It is easily checked for the case $I = 1$ by solving a set of three simultaneous differential equations for N_1, N_0, N_{-1} that the relation (2.47) is in general not satisfied at all times. We shall prove, however, that for $\delta = \gamma \hbar H_0/k \ T \ll 1$, a temperature and a relaxation time can always be defined. There is hardly a loss of generality, since the condition $\delta \ll 1$ is fulfilled down to 0.1 °K. We start from the set of $2I+1$ equations

$$\frac{d N_I}{d t} = - W_{I-1, I} N_I + W_{I, I-1} N_{I-1} \qquad (2.48)$$

$$\frac{d N_{I-1}}{d t} = (-W_{I-2, I-1} - W_{I, I-1}) N_{I-1} + W_{I-1, I} N_I + W_{I-1, I-2} N_{I-2}$$

etc. with

$$W_{m+1, m} = W (I - m)(I + m + 1) \exp(- m \ \delta), \qquad (2.49)$$

where W is the transition probability from the state $m = -{}^1/_2$ to $m = {}^1/_2$, if $I = {}^1/_2$.

From (2.49) we derive a set of linear combinations for $N_m + N_{-m}$

$$\frac{d (N_m + N_{-m})}{d t} = -W_{m+1, m} + W_{m-1, m}) (N_m + N_{-m}) +$$

$$+ W_{m, m+1} (N_{m+1} + N_{-m-1}) + W_{m, m-1} (N_{m-1} + N_{-m+1}) +$$

$$+ 2 m \ \delta N_{-m} + 2 (m+1) \ \delta N_{-m-1} + 2 (m-1) \ \delta N_{-m+1} \qquad (2.50)$$

The last three terms on the right hand side can be neglected compared to the first three.

Next we derive a set of equations for $N_m - N_{-m}$. We make use of the relation $2 m \delta N_{-m} = N_m^\infty - N_{-m}^\infty$, the difference of the population of the levels m and $-m$ at thermal equilibrium. We obtain

$$\frac{d(N_m - N_{-m})}{dt} = -(W_{m+1,m} + W_{m-1,m})[N_m - N_m^\infty - (N_{-m} - N_{-m}^\infty)] +$$

$$+ W_{m,m+1}[N_{m+1} - N_{m+1}^\infty - (N_{-m-1} - N_{-m-1}^\infty)] +$$

$$+ W_{m,m-1}[(N_{m-1} - N_{m-1}^\infty) - (N_{-m+1} - N_{-m+1}^\infty)] \quad (2.51)$$

Equations (2. 50) can be solved by anticipating that $N_m + N_{-m}$ is independent of the time, $N_m + N_{-m} = 2 N/2 I + 1$. If I is an integer, $N_0 = N/2 I + 1$ is constant in time. The solution of the equations (2. 51) is obtained by making the supposition that the difference in population between two adjacent levels $n = N_m - N_{-m}/2 m$ is independent of m. We find that the set (2. 51) reduces to á single equation

$$\frac{dn}{dt} = -2 W (n - n^\infty) \quad (2.52)$$

The result is that the relaxation time for a nucleus with spin I is the same as for a nucleus with the same magnetogyric ratio, but spin $^1/_2$.

Using (2. 40) we find for the relaxation time

$$1/T_1 = {}^3/_4 \gamma^4 \hbar^2 I(I+1) [J_2(2\nu_0) + 2J_1(\nu_0)] \quad (2.53)$$

and analogous expressions besides the equations (2. 41) and (2. 42).

2.6. The line width.

We now return to the terms A and B in (2. 34) to see what happens to the line width, when the nuclei are changing their positions as they take part in the Brownian motion. These terms will still represent secular perturbations, if we take the components near zero frequency in $J_0(\nu)$, the Fourier spectrum of $\Sigma_j (1 - 3\cos^2 \vartheta_{ij}(t))/r_{ij}^3(t)$. The question is, which frequencies must be considered to be „near zero". We may say that the perturbation is secular up to that frequency, for which $h\nu$ is of the same order as the actual splitting of the energy levels by the perturbation. The actual width expressed in cycles /sec may be called

$2 \triangle \nu = \triangle \omega / \pi = 1 / \pi T_2'$. A combination of (2. 36) and (2. 37) yields the relation for the width

$$\frac{1}{T_2'} = {}^3/_2 \, \gamma^2 \, \hbar \sqrt{\frac{I \, (I+1)}{3} \int_{-1/ \pi T_2'}^{+1/ \pi T_2'} J_0 \, (\nu) \, d\nu} \qquad (2.54)$$

If the nuclei do not all have the same magnetogyric ratio, we can alter the treatment for term A in the same way as we did in the preceding section for T_1. The term B, however, changes its character completely, because it now contains a time factor $\exp i \, (\gamma_1 - \gamma_2) \, H_0 \, t$. The components of $J_0 \, (\nu)$ at $\nu = \pm \, (\gamma_1 - \gamma_2) \, H_0 / 2 \, \pi$ determine the transition probability for a process in which to antiparallel nuclei with different magnetogyric ratios jump together. We shall not discuss it further.

The observed line width is not determined by T_2' only. There also is a contribution arising from the finite life time of the nucleus in a given state by the relaxation processes, represented by the terms C, D, E and F. For $I = {}^1/_2$, the life time of both the upper and lower level is $1/W = 2T_1$. As Weisskopf and Wigner (W 3) pointed out, we have to think that each of these levels is broadened into a continuous distribution

$$g^1 \, (\nu) \, d\nu = \frac{W}{4 \, \pi^2 \, (\nu - \nu_0)^2 + {}^1/_4 \, W^2} \, d\nu$$

The distribution in the intensity of a resonance transition between the two levels is consequently

$$g \, (\nu) = \frac{2 \, T_1 \, (\nu_0^2 + \nu^2)}{\nu^2 + 4 \, \pi^2 \, (\nu_0^2 - \nu^2)^2 \, T_1^2}$$

or to a high degree of approximation

$$g \, (\nu) \approx \frac{4 \, T_1}{1 + 16 \, \pi^2 \, T_1^2 \, (\nu - \nu_0)^2} \qquad (2.55)$$

This is the broadening caused by the finite relaxation time T_1. The distribution over the resonance frequencies, caused by the perturbation A and B, will often have an approximately Gaussian shape,

$$g^{11} \, (\nu) = T_2' \sqrt{\frac{1}{2\pi}} \, e^{-2 \, \pi^2 \, (\nu - \nu_0)^2 \, T_2'^2} \qquad (2.56)$$

Combining (2. 55) and (2. 56) we get for the line shape

$$\varphi \, (\nu) = \sqrt{\frac{8}{\pi}} \, T_1 \, T_2' \int_{-\infty}^{+\infty} \frac{e^{-2 \, \pi^2 \, (\nu^1 - \nu_0)^2 \, T_2'^2}}{1 + 16 \, \pi^2 \, (\nu^1 - \nu)^2 \, T_1^2} \, d\nu^1 \qquad (2.57)$$

Roughly we can say that the combination of two bell shaped curves will yield another bell shaped curve, of which the width is about the sum of the widths of the composing curves. We can express the total line width by

$$\frac{1}{T_2} \approx \frac{1}{T_2'} + \frac{1}{2\,T_1} \qquad (2.58)$$

Instead of the unwieldy distribution (2.57), we shall assume that $\varphi(\nu)$ can be well represented either by a Gaussian distribution like (2.56) or by a damped oscillator distribution like (2.55). This may seem

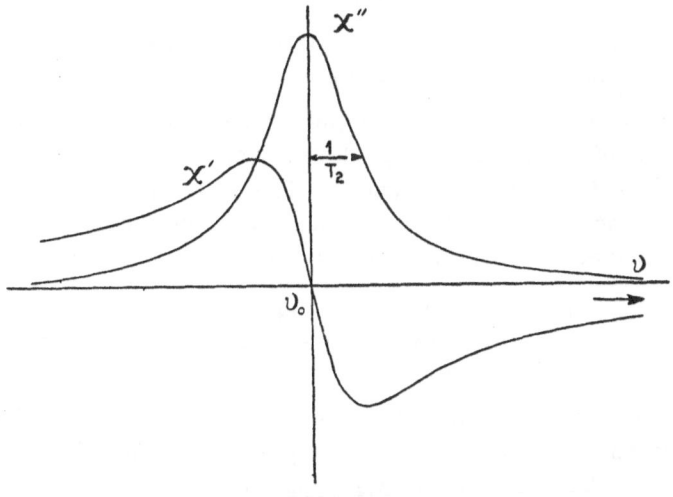

Figure 2.4.

The behaviour of the real and imaginary part of the magnetic susceptibility near resonance. The curves are drawn for the case of a damped harmonic oscillator (compare section 2.7).

arbitrary, but the choice of (2.56) was already arbitrary, and the two possibilities, which we admit, are good representatives for the case that there are practically no tails (gaussian), and that the tails of the resonance curve contribute considerably (damped oscillator). Note that in the latter case $2/T_2$ is the total width between the half maximum points. For the Gaussian $1/T_2$ denotes the root mean square deviation of the frequency. This quantity diverges for the damped oscillator curve.

We have derived general expressions for the relaxation time and line width in terms of the local field spectrum. In order to evaluate these quantities for any given substance, we only have to compute the intensity of this spectrum. In chapter 4 applications are given for solids, liquids and gases.

The meaning of the quantities T_1 and T_2' in terms of the density function $f(\nu)$ represented in fig. 2.3 might be formulated in the following manner. The time, which is required to restore the shape of $f(\nu)$ around the frequency ν_0 after the equilibrium has been disturbed in some way or other, is T_2'. The time required to restore the area under the $f(\nu)$ curve around ν_0 is T_1. We shall now see how the introduction of the quantities T_1 and T_2, which describe the interactions between the spins, changes the results obtained in the beginning of this chapter for free spins.

2.7. Classical theory with interactions.

We shall first give a brief outline of Bloch's phenomenological theory (B 4). An assembly of spins is initially so oriented in a magnetic field H_0 as to give a resultant macroscopic magnetisation $\vec{M}$. Bloch notes that the spreading of the energy levels makes the nuclei precess with slightly different Larmor frequencies. Therefore the nuclei get out of phase in a time $t \sim T_2$ and the magnetisation in the x and y-direction will be destroyed. The perpendicular components satisfy the equation

$$\frac{d M_x}{d t} = - M_x/T_2 , \quad \frac{d M_y}{d t} = - M_y/T_2 \qquad (2.59)$$

The z-component, however, will change appreciably only in a time T_1, and will reach for time $t \gg T_1$ its equilibrium value M_0.

For M_z we have the differential equation

$$\frac{d M_z}{d t} = -(M_z - M_0)/T_1 \qquad (2.60)$$

Bloch mentions that $T_1 \gg T_2$. We shall see in chapter 4 that this is not always the case, and in the following no use of this inequality is made. We can now write down the equations of motion in case of an applied r.f. field by modifying the equations (2.59) and (2.60) to

$$\frac{d M_z}{d t} = \gamma \, [\vec{H} \times \vec{M}]_z + (M_0 - M_z)/T_1$$

$$\frac{d M_{x,y}}{d t} = \gamma \, [\vec{H} \times \vec{M}]_{x,y} - M_{x,y}/T_2 \qquad (2.61)$$

To solve these equations, for a magnetic field (1.10)

$$H_x = H_1 \cos \omega t, \ H_y = H_1 \sin \omega t \text{ and } H_z = H_0$$

transform again to a rotating coordinate system

$$M_x = u \cos \omega t - v \sin \omega t$$

$$M_y = \pm (u \sin \omega t + v \cos \omega t)$$

The — sign is used for negative γ. Using the abbreviations

$$\tau = b t$$

$$b = |\gamma| H_1$$

$$a = 1/b T_1$$

$$\beta = 1/b T_2$$

$$\delta = (\omega_0 - \omega)/b = \triangle \omega/b$$

the equations of motion reduce to

$$\frac{d u}{d \tau} + \beta u + \delta v = 0$$

$$\frac{d v}{d \tau} + \beta v - \delta u + M_z = 0 \qquad (2.62)$$

$$\frac{d M_z}{d \tau} + a M_z - v = a$$

A time independent solution is possible with $\dfrac{d u}{d \tau} = \dfrac{d v}{d \tau} = \dfrac{d M_z}{d \tau} = 0$,

we obtain: $M_z = \dfrac{1 + (T_2 \triangle \omega)^2}{1 + (T_2 \triangle \omega)^2 + (\gamma H_1)^2 T_1 T_2} M_0$

$$u = \frac{|\gamma| H_1 T_2 (\triangle \omega T_2)}{1 + (T_2 \triangle \omega)^2 + (\gamma H_1)^2 T_1 T_2} M_0$$

$$\qquad (2.63)$$

$$v = \frac{- |\gamma| H_1 T_2}{1 + (T_2 \triangle \omega)^2 + (\gamma H_1)^2 T_1 T_2} M_0$$

The strength of the resonance effect we want to measure is — as will be shown in more detail in the next chapter — proportional to the components of the magnetisation perpendicular to H_0, which vary with the same frequency as the applied signal. We shall therefore maximize the expressions (2.63) for u and v with respect to the frequency ω and the applied field H_1.

We see immediately that the out-of-phase component v of the magnetisation which is responsible for absorption, is maximal for $\triangle \omega = 0$. The optimal value

$$v = {}^1/_2 \ M_0 \ \sqrt{T_2/T_1} \tag{2.64}$$

is then obtained if $\gamma^2 H_1^2 T_1 T_2 = 1$.

The component u, which is in phase with the magnetic field H_1 and describes the dispersion, reaches asymptotically the same maximum value as v for $\triangle \omega = \sqrt{1 + \gamma^2 H_1^2 T_1 T_2}/T_2$ and $\gamma^2 H_1^2 T_1 T_2 \gg 1$.

If we introduce the complex magnetic susceptibility $\chi = \chi' - i\chi''$ and write $\vec{M} = \chi \vec{H}$, then it follows that $u = \chi' H_1$, $v = \chi'' H_1$ and from (2.63)

$$\chi' = \frac{|\gamma| T_2 (\triangle \omega T_2) \chi_0 H_0}{1 + (T_2 \triangle \omega)^2 + \gamma^2 H_1^2 T_1 T_2} \tag{2.65}$$

$$\chi'' = \frac{|\gamma| T_2 \chi_0 H_0}{1 + (T_2 \triangle \omega)^2 + \gamma^2 H_1^2 T_1 T_2} \tag{2.66}$$

where $\chi_0 = M_0/H_0$ is the static susceptibility (1.18).

The susceptibility must be considered to have different values for the three components of $\vec{H}$. For the z-component of H and the component rotating in the opposite direction around the z-axis, the susceptibility has the static value χ_0; no resonance phenomenon occurs. We see from (2.63), (2.65) that in strong radio-frequency fields of the resonating component a saturation effect occurs. The susceptibilities decrease for $\gamma^2 H_1^2 T_1 T_2 \gg 1$ in proportion to H_1^{-2}. The externally induced transitions then compete too succesfully with the transitions caused by the relaxation mechanism. In the case $T_1 = \infty$, the magnetisation is zero. This is the heating up (1.24) in a system of free spins. In the limit $H_1 \to 0$, when $\gamma^2 H_1^2 T_1 T_2 \ll 1$, the line shape (2.65) is identical with a damped oscillator curve like (2.55), as shown in fig. 2.4. This is no surprise, since we have rather arbitrarily assumed the exponential decay of the perpendicular components in (2.59). The reasoning in this section cannot be considered as a proof that the line shape near resonance is the same as that of a damped harmonic oscillator. In the preceding sections we have already seen that this is in general not the case. Bloch's classical equations (2.62) are especially useful to obtain solutions for non-stationary phenomena and transient effects (B 7). For these the reader is referred to the original literature.

2. 8. *Quantumtheory with interactions.*

We shall now derive the saturation formulæ for absorption and dispersion in the stationary state along quantummechanical lines (G 6, K 8). We start with the absorption for $I = \frac{1}{2}$. The situation to be described is that of a competition between the applied signal and the local field spectrum, the former tending to make the surplus number $n = N^+ - N^-$ zero, the latter to keep it at the value of temperature equilibrium

$$n_0 = \frac{N^+ + N^-}{2} \; \gamma \, \frac{\hbar \, H_0}{k \, T}.$$

The surplus number n can formally be thought of as being distributed over a frequency range according to a function $\varphi(\nu)$, as was discussed in section 2. 6.

$$n(\nu) = n \, \varphi(\nu) \qquad \int_0^\infty \varphi(\nu) \, d\nu = 1 \qquad (2.67)$$

The frequency of the applied signal $H_1 = \int H_1(\nu) \, d\nu$ is so well defined, that $\varphi(\nu)$ can be considered as constant over the region where H_1 is different from zero. In the radio frequency range such a pure sine wave for H_1 is practically realisable. According to (2. 16) this signal causes a surplus number of transitions per second

$$\frac{1}{4} \gamma^2 H_1^2 \, n \, \varphi(\nu) \qquad (2.68)$$

In the stationary state this must be equal and opposite to the number of transitions caused by the relaxation mechanism.

$$-W(n - n_0) = -(n - n_0)/2 \, T_1 \qquad (2.69)$$

Equating expressions (2. 68) and (2. 69), we find for n in the stationary state

$$n = n_0 \, \frac{1}{1 + \frac{1}{2} \, T_1 \, H_1^2 \, \gamma^2 \, \varphi(\nu)} \qquad (2.70)$$

The power absorbed in the stationary state is obtained by substituting (2. 70) in the expression (2. 68) and multiplying by the energy $h\nu = \gamma \hbar H_0$ absorbed in each transition. Here we assume that the shape of the distribution of the surplus number is independent of H_1.

$$P = \frac{\frac{1}{4} \, n_0 \, \varphi(\nu) \, \gamma^2 \, H_1^2 \, \gamma \hbar \, H_0}{1 + \frac{1}{2} \, \gamma^2 \, H_1^2 \, T_1 \, \varphi(\nu)} \qquad (2.71)$$

The absorbed power can be expressed in terms of the complex susceptibility. The energy of the system is

$$-\int \vec{M} \cdot d\vec{H} = -\int \vec{M} \cdot \frac{d\vec{H}}{dt} dt.$$

The power absorbed by it,

$$P = \frac{-d}{dt} \int \vec{M} \cdot \frac{d\vec{H}}{dt} dt = -\vec{M} \cdot \vec{H},$$

is made up from the contributions of v in the x- and y-component of M. With the use of $v = \chi'' H_1$ the result for the average absorbed power is

$$P = \omega \chi'' H_1^2 \tag{2.73}$$

From (2.71) and (2.73) we find

$$\chi'' = {}^1/_4 n_0 \gamma^2 \hbar \frac{\varphi(\nu)}{1 + {}^1/_2 \gamma^2 H_1^2 T_1 \varphi(\nu)} \tag{2.74}$$

Using the relation $M = {}^1/_2 n_0 \gamma \hbar$, $\omega \approx \gamma H_0$ near resonance, and taking for $\varphi(\nu)$ a distribution (2.55) we get back to Bloch's expression (2.66), as it should be.

If $I \neq {}^1/_2$, we must have a detailed balance for transitions between each pair of adjacent levels. The transitions probabilities for the applied signal and the relaxation mechanism depend in the same way on m_I. With the result of (2.52) the difference in population between any two adjacent levels is again given by (2.70), where now $n_0 = \frac{N}{2I+1} \frac{\gamma \hbar H_0}{kT}$. Already in (1.16) we have seen that the result of summing over all values of m leads to the result, that the power absorbed is proportional to $I(I+1)$. So we find immediately

$$\chi'' = {}^1/_3 N H_0 \frac{\gamma^3 \hbar^2}{kT} \frac{\varphi(\nu)}{1 + {}^1/_2 \gamma^2 H_1^2 T_1 \varphi(\nu)} I(I+1) \tag{2.75}$$

The spin I can be determined from the total line intensity.

The corresponding formulæ for the dispersion can be derived by a general theorem. Kramers (K 7, K 9) showed that the relation

$$\chi'(\nu) = \frac{2}{\pi} \int_0^\infty \frac{\nu' \chi''(\nu')}{\nu'^2 - \nu^2} d\nu' \tag{2.76}$$

exists between the real and imaginary part of an electric or magnetic susceptibility. But if the absorption (2.74) is described by Bloch's expression (2.65) for χ'' also the corresponding dispersion must be described by the expression (2.66) for χ'. The integration (2.76) must give this result (2.66).

We have not emphasized a very important assumption in the derivation of the saturation formulæ. The distribution function $\varphi(\nu)$ has been taken independent of H_1, that is: the shape of the distribution of the surplus nuclei is independent of the saturation. In fig. 2.5. the ori-

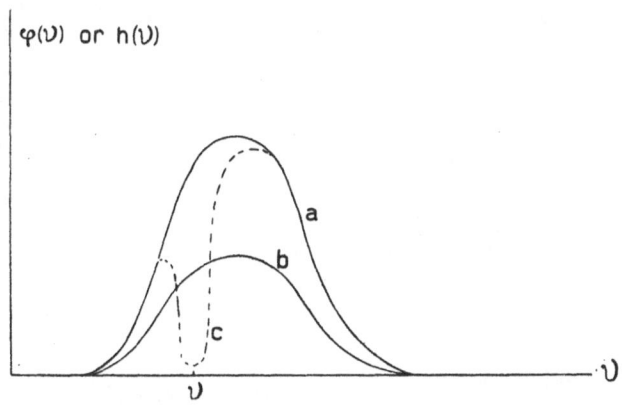

Figure 2.5.

The distribution of the surplus population over the resonance frequencies.
a) No saturation
b) Saturation without change of shape
c) Partial saturation in inhomogeneous fields.

ginal distribution is represented by curve a, b is a saturated distribution of the same shape, and will be realised, if $T_1 \gg 1/\gamma H_1 \gg T_2$. Then the rapid spin-spin interaction will be able to maintain the distribution b, although the signal only induces transitions near the frequency ν. The assumption is still approximately true if $T_1 \gg 1/\gamma H_1$ and $T_2 \gg 1/\gamma H_1$. For then the large applied field will produce transitions over the whole width of the resonance curve. But in this case formula (2.65) is not quite correct as follows from the consideration, that in the limit of very large H_1, we must get Rabi's formula (2.11) with a line width $\triangle \omega = \gamma H_1$, while (2.65) gives

$$\triangle \omega = \gamma H_1 \sqrt{T_1/T_2}$$

One might infer from this that the saturation is still rather well described by (2.71), if T_1 is of the same order as T_2. In chapter 4 we shall see that this is the case for liquids ands gases. For a special model of He-gas the Bloch formula will be derived in a independent way. The assumption that the shape of the distribution function does not change is definitely wrong, if the distribution is mainly determined by inhomogeneities $\triangle H$ in the field H_0. In this case the saturated distribution will look like curve c in fig. 2.5. The nuclei which are in another part of the field are not at resonance and there the unsaturated distribution $h(\nu)$ remains. The line shape, determined by a point-by-point measurement at various frequencies ν of the applied signal, is always $h(\nu)$, independent of the degree of saturation. We have in general:

$$\chi'' = \gamma M_0 \int_{-0}^{+\infty} \frac{T_2}{1 + T_2^2 (\omega - \omega')^2 + \gamma^2 H_1^2 T_1 T_2} \, h(\nu') d\nu' \quad (2.77)$$

If $h(\nu)$ varies only slowly in the region where transitions are appreciable this becomes

$$\chi'' = h(\nu) \gamma M_0 \int_{-\infty}^{+\infty} \frac{T_2}{1 + (T_2 \triangle \omega')^2 + \gamma^2 H_1^2 T_1 T_2} \, d\nu'$$

$$\tag{2.78}$$

$$= h(\nu) \frac{\gamma M_0}{2\sqrt{1 + \gamma^2 H_1^2 T_1 T_2}}$$

Finally it must be stressed that all formulæ are derived for a rotating magnetic field. Experimentally always an oscillating field is used. This, however, can be decomposed into two fields rotating in opposite directions. The field $H_x = 2 H_1 \cos \omega t$. $H_y = 0$ is equivalent to the set

$$H_x = H_1 \cos \omega t \qquad H_x = H_1 \cos \omega t$$

$$\tag{2.79}$$

$$H_y = H_1 \sin \omega t \qquad H_y = - H_1 \sin \omega t$$

Now only one of these — the right or left circular polarised one for a positive or negative magnetogyric ratio respectively — will be effective in producing transitions. The transition probability, the absorbed power, the components of magnetisation u and v etc. in a linear oscillating field with amplitude $2 H_1$ are the same as in the rotating field with a component H_1 perpendicular to H_0, for which all discussions were made. If one prefers to express the magnetisation u and v in terms of the linear field H_1, one must use a susceptibility χ_{lin}, which is one half of that occuring in (2.65). Note that H_1^2 in the saturation term of that formula must be replaced by $^1/_4 H_1^2$. The maximum value of the magnetisation χH_1 remains unchanged:

$$u_{max} = v_{max} = {}^1/_2 \, M_0 \sqrt{T_2/T_1}$$

The reader can easily adapt the other formulæ to the case of a linear field H_1. Bloch and Siegert (B 9) have discussed the influence of the other rotating component, which we neglected. The resonance frequency is displaced by $(H_1/H_0)^2 \, \nu_0$ in order of magnitude. Under experimental conditions one always has $H_1/H_0 < 10^{-3}$. So this correction is completely negligible. The only disadvantage of a linear field is that the sign of the magnetogyric ratio cannot be determined. Nevertheless this information has been obtained from resonance experiments, for which we must refer to the literature (M 3).

CHAPTER III

THE EXPERIMENTAL METHOD.

3. 1. *The experimental arrangement.*
3. 1. 1. *The original experiment, by Purcell, Torrey and Pound.*

From the preceding chapters it should be clear that the problem is to measure a small change of the magnetic susceptibility in the radio-frequency range, caused by the resonance of the nuclei. The essential part of the apparatus will therefore be a coil, placed in a constant magnetic field H_0 and filled with a material, which contains the nuclei to be investigated. The coil is tuned by a condenser in parallel, and a current ·in the coil is excited by coupling it to a generator. The position of the coil is such, that the radio-frequency field in the coil is perpendicular to H_0. On the other hand the circuit is coupled to the input of a receiver, tuned to the same frequency. At the output of the receiver we measure the transmission through the LC-circuit. We vary H_0 slowly. On hitting the resonance of the nuclei at the value $H_0 = 2\pi\nu_0/\gamma$, a change in the output reading will be recorded, as the absorption by the nuclei lowers the Q of the LC-circuit by a small amount. The quality factor Q is defined by

$$Q = 2\pi \, \frac{\text{energy stored in the circuit}}{\text{power dissipated per cycle}} \qquad (3.\,1).$$

At resonance the power absorbed per unit time is increased by the energy absorption of the system of nuclei. The real part of the nuclear susceptibility causes a shift in phase of the radio-frequency signal, which was not detected by the original arrangement. It is plausible, and will be proved in section 2 of this chapter, that the relative decrease in voltage across the coil is proportional to the change in the nuclear susceptibility and the Q of the coil. In the original experiment by Purcell, Torrey and Pound, the resonant circuit consisted of a short length of coaxial line, tuned by a capacitance to 30 Mc/sec as shown in fig. 3. 1. The power was inductively coupled in and out by small loops A and B. The inductive part of the cavity was filled with 850 cc. paraffin. The Q obtained was 670. Assuming a line width of 5 oersted in paraffin, a change in χ of the order of 3.10^{-6} could be expected, corresponding to a change of less than 1 $\%$ in output power.

To increase the relative change a bridge circuit was built, which balanced out the main part of the signal going into the receiver. By means of an attenuator and pieces of coaxial line of variable length the signal going through one branch could be adjusted so as to have about the same amplitude and 180° phase shift with respect to the signal going through the cavity in the other branch of the bridge.

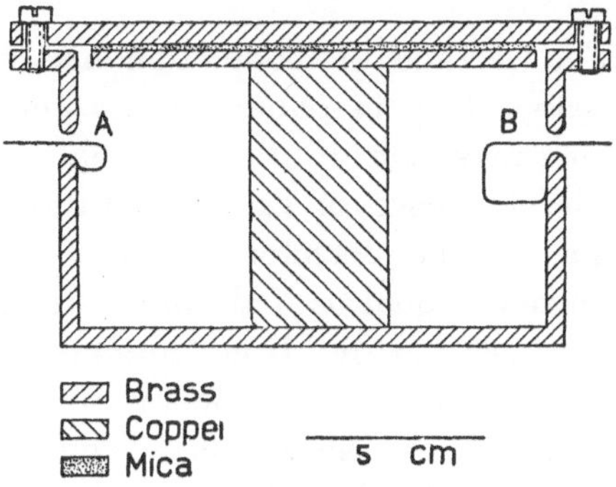

Figure 3. 1.

Cavity, used in the first experiment of nuclear magnetic resonance. The cavity was filled with 850 cc. Paraffin.

At the resonance of the protons in the paraffin the output changed by 50 $\%$. The change was positive or negative depending on the way the bridge was balanced. The resonance occurred at that value of the magnetic field H_0, which could be expected from Rabi's measurement of the proton moment and the resonance frequency used. With the same circuit also the resonance of H^1 and F^{19} nuclei were observed in a mixture of $Ca F_2$ powder and mineral oil. The amplitude of the radio-frequency field in the cavity was kept as low as possible ($< 10^{-4}$ oersted) in order to avoid saturation. From (2. 65) we see that saturation occurs if $\gamma^2 H_1^2 T_1 T_2 \approx 1$.

Assuming $T_2 = 10^{-5}$ sec, corresponding to a width of the proton line of about 3 oersted, we find that the relaxation time could be about four hours without saturating the spin system. An upper limit for the relaxation time in paraffin of 60 sec was established by observing the resonance as quickly as possible after the magnetic field H_0 is applied. Immediately after the field is switched on, the relative population of the protons in the upper and lower level will be given by the a-priori probabilities of these levels, which are equal. Therefore a resonance can only be observed after a time comparable to the relaxation time has elapsed. This upper limit of 60 sec is historically important, since the available information at that time, consisting of Waller's theory and Gorter's experiments, indicated very long relaxation times.

3. 1. 2. *Further experimental development.*

Already in their first papers Purcell, Torrey and Pound (P 3) mentioned the advantage of modulating the magnetic field H_0 with an alternating field of small amplitude and low frequency,

$$H_z = H_0 + H_s \sin \omega_s t \tag{3.2}$$

The method of modulation had been applied in other fields on many occasions (e.g. in photo-electric amplifiers (M2)). The advantage is not an improvement in the essential signal-to-noise ratio, but a reduction of the influence of external disturbances. Instead of measuring the d-c output of the detector directly, in the modulation method the audio-frequency signal is fed, after detection, into an audio-amplifier which has a very narrow band around the frequency ω_s, which was taken equal to half the frequency of the mains, i.e. 30 cycles per second. One could use an ordinary audio-amplifier with an a.c. galvanometer at the output. Actually a phase-sensitive „lock-in" amplifier (D3) was used, which will be described in the next section. The cavity was replaced by a coil, of about 2 cm long and 0.7 cm in diameter, tuned by a variable condensor. The quality factor was about 10 times smaller for this circuit compared to the preceding one, corresponding to a similar decrease in linear dimensions. This implies, of course, a loss in signal strength. The cavity, however, required a large magnet, which was not permanently available. Furthermore even in the large magnet the field was not homogeneous over the region of the cavity, so that the maximum signal from the nuclei was decreased according to the lower value of h(ν) in (2.78). In addition the cavity required unwieldy and sometimes not available quantities of material. Also the bridge circuit was improved as described in the next section.

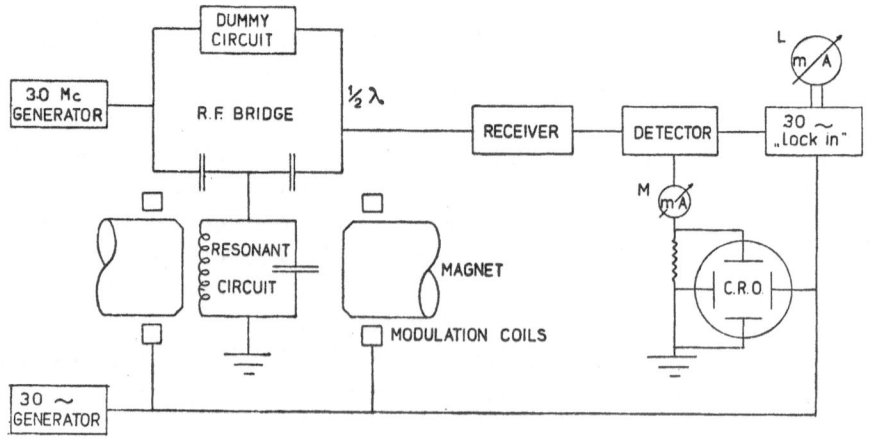

Figure 3. 2.

Block diagram of the experimental arrangement, which is described in sections 3.1.3. and 3.1.4.

With the apparatus represented schematically in fig. 3. 2 the nuclear resonance could be observed either on the screen of the oscilloscope or on the output L of the 30 ~ audio amplifier. In the first case the amplitude of the 30 ~ modulation sweep is larger than the line width. On passing through the line the detector current measured by M changes proportional to the absorption by the nuclei, if the bridge is balanced properly. The change in voltage across a resistor in the detector circuit is put directly on the vertically deflecting plates of the cathode ray tube. On the horizontal plates we put a synchronous 30 ~ sine sweep. On the screen appear two absorption curves, because in each cycle the sweep passes twice through resonance. The two curves can be made to coincide by proper adjustment of the phase of the horizontal sweep (see fig 3. 15). By changing the main field H_0 the absorption lines would shift together and finally disappear from the screen. A section of the magnetic field from $H_0 - H_s$ to $H_0 + H_s$ is plotted in the horizontal direction on the oscilloscope. For observation on the meter L the sweep amplitude was made small compared to the line width. As will be shown later the deflection of this meter was then at any point H_0 proportional to the derivative of the curve on the oscilloscope at the corresponding field strength. By changing H_0 very slowly the entire derivative function could be measured point by point. We shall now describe the apparatus in the block diagram in more detail.

3. 1. 3. *The radiofrequency bridge.*

It is important to be able to balance the bridge independently in amplitude and phase. To obtain these two orthogonal adjustments first the circuit shown in fig. 3. 3. was used. The various parts are linked together with coaxial lines, of which the outer conductors are

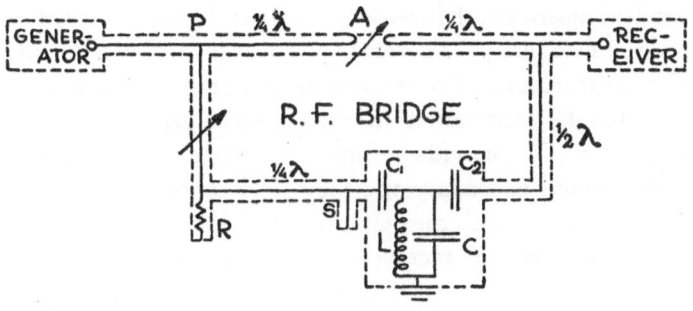

Figure 3. 3.

Radio-frequency bridge, described in the text.
$C_1 = 3 \, \mu\mu\text{F}$ $C_2 = 5 \, \mu\mu\text{F}$ $C = 100 \, \mu\mu\text{F}$.

connected to the grounded terminals of the generator, resonant circuit and receiver. The attenuator A was of the inductive type, used for 10 cm. waves. The inner conductors of the coaxial lines were terminated in small loops, the distance between which could be varied. The attenuation between the loops is characteristic for a wave guide (K 11) „beyond cut-off". At 30 Mc/sec the impedance of the loops is practically zero. The attenuation is large and not accompanied by a noticeable phase shift. In order to obtain the same low transmission in the other branch a very short closed stub S was inserted, which also acted practically like a short circuit. To this voltage generator of very low internal impedance the resonance circuit was loosely coupled by a small condenser C_1, while on the other side it was critically coupled to the receiver input by another condenser C_2. The non-resonant line between P and the carbon resistor R, which is equal to the characteristic impedance of the coaxial cable is of variable length and provides a pure phase adjustment. The lines of one quarter wave length transform the short circuits into open circuits. Looking from the generator into the bridge one „sees" the resistance R. This expression means that the impedance between the two terminals of the bridge at the side of the generator is R. Between the terminals, which are connected to the input of the receiver, is the impedance of the LC circuit, tranformed by C_2. If this coupling has the desired critical value, the maximum available power from the resonance circuit will flow into the receiver. The disadvantages of this bridge are its low transmission and its asymmetry. The minimum attenuation is about 40 db, and the generator has to be unnecessarily powerful. The power required for the nuclear experiments is essentially low, but since the maximum output of the available generator was 2V. into 75 Ω, it was desirable not to waste too much power.

The asymmetry of the bridge allowed a balance of the bridge only at one frequency. Since the generator showed some frequency modulation, this was very undesirable. Moreover was the balance sensitive to very slow drifts of the frequency. Therefore the circuit of fig. 3. 4 was adopted for the final measurements. The generator power is fed into two coaxial lines of equal length terminated in their characteristic impedance. The condensers C_1 and C_2 provide a loose coupling to the resonance circuits, which are tuned at exactly the same frequency and have the same Q. The output condensers C_3 and C_4 provide critical coupling between the resonance circuits and the receiver, for which the two circuits appear in parallel.

Since part of the change in power from the nuclear resonance now flows into the dummy circuit instead of into the receiver, the available signal-to-noise power ratio is only one half of the optimal value, obtained

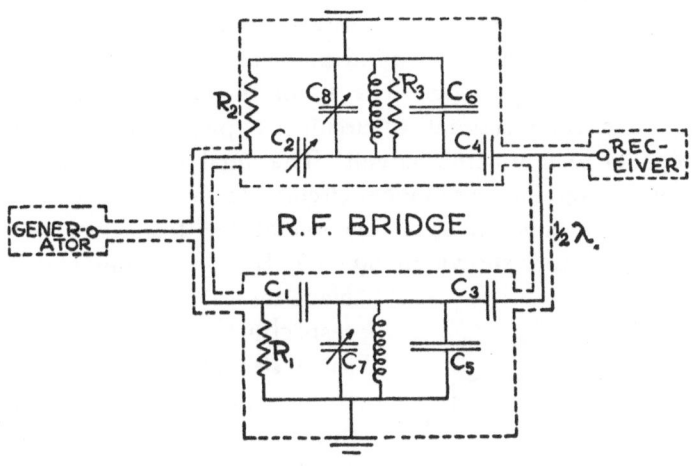

Figure 3. 4.

Improved radio-frequency bridge, described in the text.
$R_1 = R_2 = 50\ \Omega$
$C_1 = 3\ \mu\mu\,F,\ C_2 = C_7 = C_8 = 5\ \mu\mu\,F,\ C_3 = C_4 = 4\ \mu\mu\,F$
$C_5 = C_6 = 60\ \mu\mu\,F.$

with the preceding diagram. A coaxial line of one half wave length produces a phase shift of 180° degrees. For reasons to be discussed in the next section, the balance was never made complete. When the remaining unbalance is in amplitude, the imaginary part of the nuclear susceptibility is measured; when the remainder is phase unbalance, the phase shift of the signal and thus the real part is measured. The unbalance used in the experiments varied from 0.1 to 0.001 in amplitude of the original signal (20-60 db) and was stable at these values. For short time intervals balance to one part in 30000 (90 db) could be obtained. The ratio by which the signal is reduced in the bridge we shall call b. The radiofrequency coil, which is placed in the magnetic field, is shown in fig. 3. 8. The coil consists in a typical example of 11 turns of copper wire of 0.1 cm diameter. The length of the coil is 1.7 cm, and its inside diameter 0.6 cm. Samples, contained in thin walled cylindrical glass vessels of approximately 0.5 cc, could easily be placed in the coil and replaced. The brass box, which was below the magnet gap, contained the other circuit elements of the nuclear resonance circuit in fig. 3. 4.

With the trimmer C_7 minor adjustments in the phase could be made. The dummy circuit contained essentially the same elements, although no particular care was taken to make a geometrical copy. For the dummy the resonance frequency was adjusted by selecting the tuning capacitance C_6; the coupling condenser C_2 formed together with C_8 a differential air trimmer. If C_2 increased, C_8 decreased by about the same amount and this provided a phase-independent adjustment of the amplitude for the balance of the bridge. The Q of the dummy was made equal to that of the nuclear resonance circuit by putting a carbon resistor R_8 as a parallel load on the circuit. The analysis and trimming was done by connecting the resonance circuits to a signal generator and measuring the transmission as a function of the frequency with a sylvania crystal detector, shown in fig. 3.5. Resonance circuits were also made for 14.4 Mc/sec and 4.85 Mc/sec respectively. As a matter of fact, the same size of coil was used and only the condensers were adjusted. Of course also the half wave length cable had to be changed, which at the lowest frequency had a length of about twenty meters. The Amphenol cables have a polythene dielectric and a characteristic impedance of 50 Ω. It might be worth while to obtain the 180° phase shift by a radio frequency transformer with centre-grounded secondary instead of a difference in line length, although the method used by us proved to be successful.

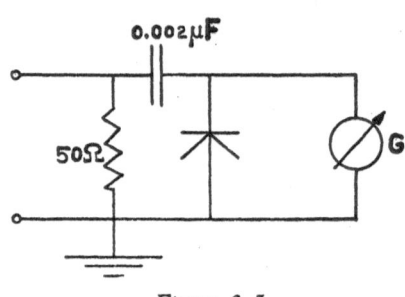

Figure 3. 5.

Crystal detector with galvanometer.

3. 1. 4. *Other apparatus.*

The signal generator was of the type General Radio 805 C. The frequency ranged from 16 kc to 50 Mc. The output could be varied continuously from 0.1 μ V to 2 V. The characteristic output impedance was 75 Ω. So there was a mismatch of three to one with the bridge. This was harmless and no attempt was made to eliminate it, as could be done, e.g. by a quarter wave transformer. The receiver is a National H R O 5. The narrow passband of this instrument (1500 cycles/second) was of little use in the present experiment. It necessitated more frequent adjustments of the tuning. Only for observation with the oscilloscope it improved the signal to noise ratio. The H R O 5 was fed by an electronically regulated power supply. In the experiments at 30 Mc/sec the receiver was preceded by a pre- amplifier, developed at the Radiation Laboratory, Cambridge (Mass.). Originally designed for use in Radar equipment, it had a pass band from 25-35 Mc/sec. Its use for our investigation was by virtue of its extremely low noise figure, at least 10 db less than the commercial receiver. Thus it made possible the detection of signals three times weaker in amplitude, and increased

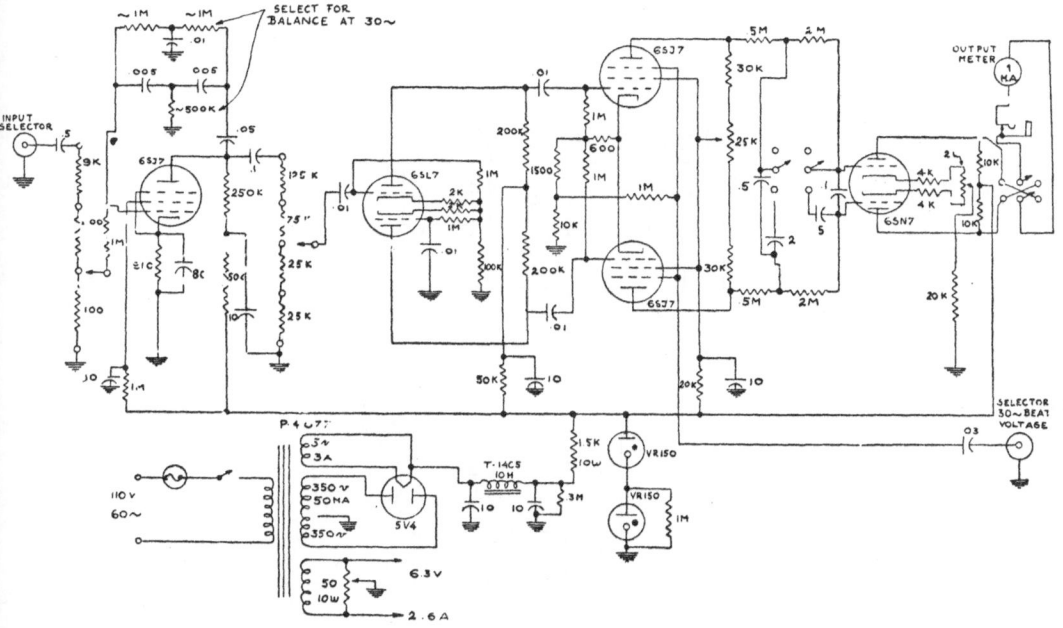

Figure 3. 6.
The phase sensitive, 30 ~ mixer-amplifier, according to Dicke (D 3).

the accuracy of measurement of signals of the same size by a factor three. The wiring diagram of the tuned phase sensitive audio-amplifier is given in fig. 3.6. The twin T feed-back filter-stage tuned at 30~, is to reduce the influence of harmonics and to prevent overload of the later stages by spurious induction from power lines etc. In the other 6 SJ 7 tubes the 30~ signals originating from the nuclei in the modulated field is mixed with a 30~ signal of about 30 V put on the suppressor grid. The last stage is a balanced amplifier for direct current. If the output meter is fast, the time constant and the pass band are essentially determined by the RC-circuits between the mixer and final stage. These lead to a differential equation for the output-reading which is similar to that of a critically damped galvanometer. The behaviour of the phase sensitive amplifier with respect to the signal-to-noise ratio is the same as for a critically damped a.c. galvanometer with time constant RC. The detection is sensitive both to frequency and phase. The generator providing the 30~ signal is represented in fig, 3.7. It consists of a multivibrator with subsequent filters to get a 30~ harmonic oscillation, a power amplifier to provide the current for the modulation coils, a phase shift circuit and an amplification stage for the beat voltage on the suppressor grids of the audio mixer, at the same time serving for the horizontal sweep of the oscilloscope. Alter-

60

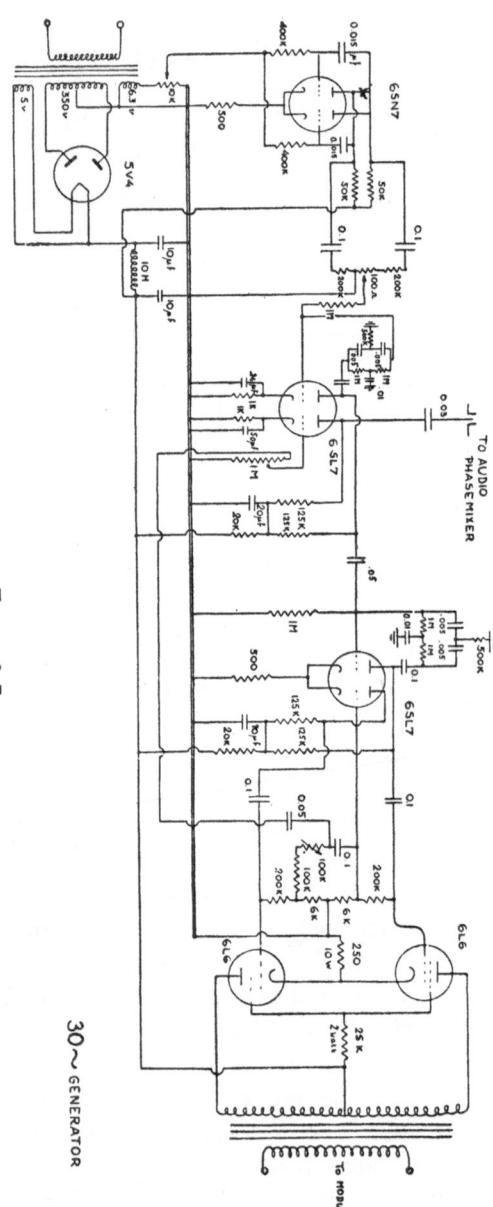

Figure 3.7.

The 30 ~ generator.

natively to the multivibrator a synchronous motor was used to drive a small $30\sim$ generator (40 Volt, 1000 Ω). The use of $30\sim$ excluded any response of the audio-amplifier to spurious $60\sim$ signals or harmonics thereof. Because the multivibrator was locked to the frequency of the mains, possible zero drifts, arising from slow changes in phase with respect to the mains, were excluded.

The magnet was made by the Société Génévoise. The pole pieces are schematically drawn in fig. 3. 8. The face of the pole pieces was 14 cm in diameter and the width of the gap was varied between 1.8 and 3 cm. For the first width a field of 7000 oersted was obtained at 15 amp. and 10 Volt, 11.000 oersted at 30 amp. and 16.000 oersted at 80 amp. The strongest field used was 8700 oersted. The current was supplied by two heavy duty truck batteries of 300 ampere hours each. An unsuccessful attempt had been made to regulate

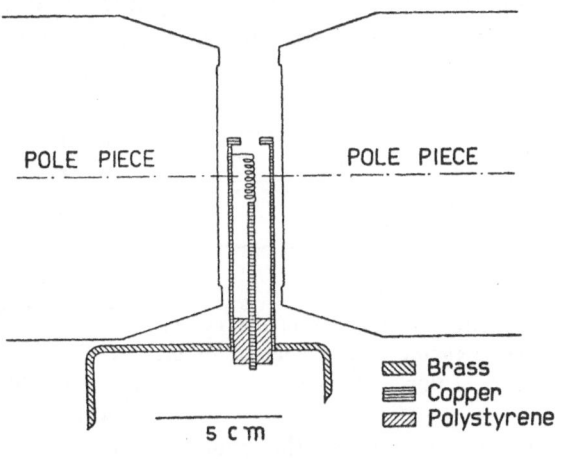

Figure 3. 8.

The radiofrequency coil, in which the nuclear resonance takes place, between the pole pieces of the magnet.

the magnetic field, if the power was supplied by a generator. Since the time constant of both the magnet and the generator were of the order of one second and at the same time and unusual large periodic perturbation in the generator voltage occurred at about this frequency caused by an imperfection in the generator, the system would break into oscillations before sufficient regulation was obtained. The batteries, however, supplied a current which was very stable over intervals of ten minutes or more. Around the pole pieces two modulation coils were placed consisting of about 50 turns each. At 0.5 amp. they could provide a $30\sim$ sweep of about 3 oersted in amplitude (6 oersted total width). The amplitude of this sweep was constant within 2% over the whole region of the gap. The main field was supposed to be made more homogeneous by the rim of 0.04 cm high and 0.8 cm wide on the edge of the pole pieces (R8). A discussion of the actual homogeneity will be made in section 3. 5. The current through the magnet coils was measured by the voltage drop across a shunt of about $10^{-2}\,\Omega$. The

E.M.F. was balanced by a Leeds and Northrup type K potentiometer. The magnet current could be adjusted roughly by the number of 2 V cells used. Then the current passed through a manganin band of $0.2\,\Omega$, of which any part could be shorted by a sliding contact. The fine regulation was achieved by a rheostat of $5\,\Omega$, in parallel to a $0.05\,\Omega$ manganin resistance in the magnet circuit.

3. 2. *The radio signal caused by nuclear resonance.*

We shall assume that the amplitude and frequency of the modulation sweep are so small, that the solutions obtained in chapter 2 for a time independent field H_θ remain valid. In section 3. 7 we shall indicate some violations of the results, if the rate of change of the magnetic field is too fast. Since the condensor C_1 in fig. 3. 4 provides a loose coupling, we can consider the resonance circuit, which contains the nuclei, as being driven by a constant current generator $i_0 = i\,\omega\,C_1\,V_1$, where V_1 is the voltage on R_1 (fig. 3. 4). The admittance at resonance of the circuit consists only of the conductance $1/R_0 = 1/\omega\,L\,Q$. At nuclear resonance the value of L changes according to

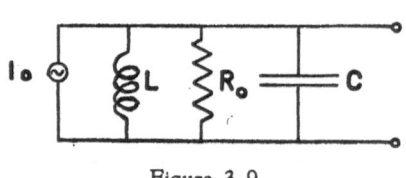

Figure 3. 9.

The aequivalent diagram of the nuclear resonance circuit, loosely coupled to the signal generator.

$$\triangle L = [1 + 4\,\pi\,(\chi' - i\,\chi'')q - (1 + 4\,\pi\,\chi_0\,q)]\,L_{vac}$$
$$\triangle L \approx 4\,\pi\,(\chi' - i\,\chi'')\,q\,L \qquad (3. 3)$$

since χ_0 is negligibly small compared to χ' and χ''. The susceptibilities are given by (2. 75) and (2. 76), and q is a filling factor. If the field H_1 in the coil were homogeneous, it would be the fraction of the volume filled by the sample. The change in admittance produces a change in voltage across the coil

$$\triangle V = V - V_0$$
$$V_0 = i_0\,R_0$$
$$V = i_0\,(\frac{1}{i\,\omega\,L} + \frac{1}{R_0} + i\,\omega\,C)^{-1}$$

Making use of (3. 3), of the expression for R_0 and of the fact that the change in admittance by the nuclear absorption and dispersion is small, we obtain finally:

$$\triangle V = \left(\frac{dV}{dL}\right)_{res} \triangle L = 4\pi q \, Q\, (\chi'' + i\chi') \, V_0 \qquad (3.4)$$

We are allowed to consider the nuclei as a voltage generator represented in fig. 3.10. The diagram holds also for the bridge of fig. 3.3. The bridge of fig. 3.4 behaves as a voltage generator $\triangle V/2$ with an internal resistance $R_0/2$.

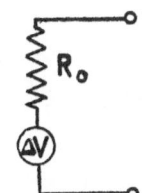

By virtue of its dependance on the susceptibility is $\triangle V$ a function of the Larmor frequency and so of the magnetic field H_0. One can plot $\triangle V$ in the complex plane. If one uses Bloch's expressions (2.65) with a factor $^1/_2$ for the linear χ' and χ'', one finds that the locus of $\triangle V$ is a circle in the case $\gamma^2 H_1^2 T_1 T_2 \ll 1$.

On substitution of $z = \triangle\omega T_2$ we have in this case

Figure 3.10.

The nuclear magnetic resonance is aequivalent to a voltage generator with internal impedance R_0.

$$(\triangle\,V)_{abs} = 2\pi q\,Q\,V_0\,M_0\,\gamma\,T_2\,(1+z^2)^{-1} \qquad (3.5)$$

$$(\triangle\,V)_{disp} = 2\pi q\,Q\,V_0\,M_0\,\gamma\,T_2\,z\,(1+z^2)^{-1} \qquad (3.6)$$

If the saturation term is taken into account, on elimination of z the locus is found to be an ellipse the eccentricity of which depends on H_1. Now it has to be remembered, that the receiver detects only changes in amplitude of the input voltage. If the balance were complete, the output of a square law detector would be proportional to $|\triangle V|^2$. Under operation conditions the balance is never made complete and the situation is represented by fig. 3.11. OA is the unbalanced signal, BO is the signal coming through the dummy circuit. If we go through the nuclear resonance by varying H_0, A goes around the circle, and the input signal of the receiver for a given value of H_0 is BC, and on the screen of the oscilloscope we see the variation of $|BC|^2$ during the sweep of the magnetic field. It has to be borne in mind that the course of C over the circle during the sweep is far from linear in H_0. The entire lower half e.g. represents the narrow part between the maximum and minimum of the dispersion curve, and C will travel much faster here than on the upper parts of the circle. On observation with the phase sensitive audio amplifier the small modulation of the field causes C to go back and forth over a small part of the circle. The amplitude of BC is then a function with a period of 1/30 sec. The amplifier is not sensitive to the harmonics and will record only the $30\sim$ component in the signal. In principle this can be calculated from fig. 3.11 for any kind of balance, that is for any point B, and for any value of H_0 and H_s. Although fig. 3.11 is drawn for the special case that the locus of $\triangle V$ is a circle, this is not essential either and the same reasoning holds for any distribution function $\varphi(\nu)$, and any value of H_1 in (2.75). Here we shall discuss the limiting case, very well approximated in practice, that the change in voltage produced by nuclear resonance AC is small compared to BA, the signal which is left after the main part of $OA = V_0$ has been balanced by OB. The relation $|AC| << |BA|$ can be written as

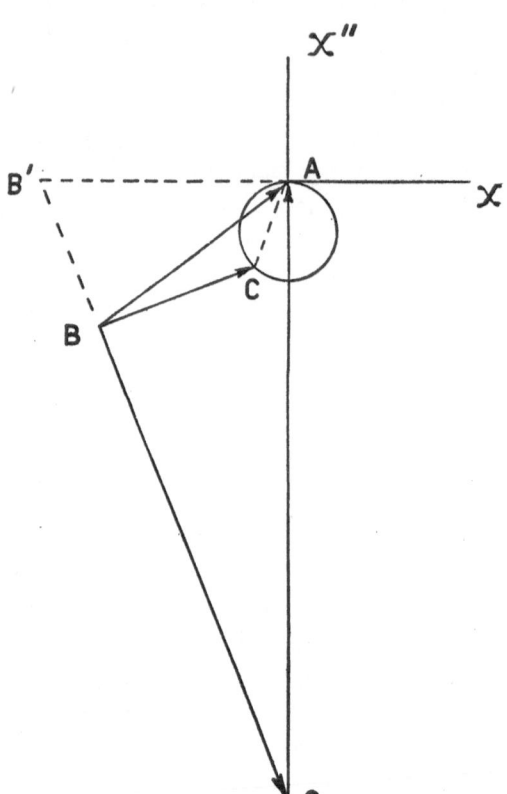

Figure 3.11.

Diagram of the voltages in two branches of the bridge, OA and BO, and of the voltage AC, generated by the nuclear resonance.

$$\triangle V << b\, V_0 \quad (3.7)$$

Assuming a square law detector, the output will be proportional to

$$|V|^2 = b^2 |V_0|^2 + \\ + 2b\,\mathrm{Re}\,(V_0 \triangle V) + |\triangle V|^2 \quad (3.8)$$

Neglecting the last term, the nuclear signal is proportional to the real part of the product $V_0 \triangle V$. If B therefore lies in line with A and O, only the real part of $\triangle V$, that is only χ'' contributes to the signal. With unbalance in amplitude the absorption is measured. If B is in the position B', only χ' contributes. With unbalance in phase the dispersion is measured. To obtain the reading of the meter L, we have to look for the $30 \sim$ component in (3.8). Assuming that the amplitude of the sweep is small compared to the line width $H_s T_2 << 1$, we can expand $\triangle V$, which via χ' and χ'' is a function of $H = H_0 + H_s \cos \omega_s t$, in a Taylor series.

$$\varDelta V(H) = \triangle V(H_0) + \frac{d(\triangle V)}{dH} H_s \cos \omega_s t + \tfrac{1}{2} \frac{d^2(\triangle V)}{dH^2} H_s^2 \cos^2 \omega_s t + \dots \quad (3.9)$$

The quadratic term has no components at ω_s. We neglect terms in $H_s{}^3$ and higher orders. It is readily seen from (3.4) and (3.9), that the voltage from the bridge can be described by a $30 \sim$ modulated wave

$$V_s = b \, V_0 \cos \omega_0 t \, [1 + m \cos \omega_s t] \qquad (3.10)$$

with

$$m = 4 \pi q \, Q \frac{d \, \chi''}{d \, H} H_s / b \qquad (3.11)$$

Formula (3.11) holds for unbalance in amplitude. On the output meter L we then measure the derivative of the absorption function. For phase unbalance we measure the derivative of the dispersion function, and a linear combination of these derivatives, if the balance is of an intermediate type.

3. 3. *Limitation of the accuracy by noise.*

In this experiment an essential limit for the obtainable accuracy is set by the thermodynamically determined fluctuations in voltage across the equivalent resistance of the generator of fig. 3. 10. To a resistance R_0 of temperature T an effective noise voltage generator has to be attributed. The mean square voltage of this generator in a frequency range $\triangle \nu$ is

$$V_n{}^2 = I_N (\nu) \, d \nu = 4 \, R \, k \, T \triangle \nu \qquad (3\,12)$$

The precision of a measurement is limited by this fluctuating voltage and the decisive quantity is the signal to noise ratio $\sqrt{V_s{}^2 / V_N{}^2}$. We shall now see what becomes of this ratio with the various methods of observation. We first suppose that the receiver has no sources of noise in itself. The voltage from the bridge can be represented by $V_s + V_n$, where V_s is given by (3.4) or by (3.10), and

$$V_n = \int_0^\infty A \, (\omega) \, e^{i \, \omega \, t} \, d \, \omega$$

$$| V_n |^2 = \int_0^\infty I_N (\nu) \, d \nu \qquad (3.13)$$

The problem consists in finding the low frequency part of the output spectrum after detection. We shall give the results for a square law detector in the case that the pass band of the receiver is a rectangle (outside the frequency range $\nu_0 - {}^1/_2 \beta, \ \nu_0 + {}^1/_2 \beta$ the gain is zero). A detailed theory of the noise with special attention to the noise in non-linear devices has been given by Rice (R 4). He also gives the results for a linear detector, which are not essentially different from the following ones. First we deal with the case that the signal is given by

$$V_s = (V_0 + \triangle V) \cos \omega_0 t \qquad (3.14)$$

and the observation is made on the screen of the oscilloscope, which has a pass band larger than β, or with the meter M (see fig. 3. 2), which we assume to have a rectangular pass band β_1. The time of indication of the instrument is related to β_1 and approximately $\tau_1 = {}^1/\beta_1$. If the balance of the bridge is complete and the signal power small compared to the noise power: $b V_0 = 0$, $\triangle V^2 << 4 R k T \beta$, the signal to noise ratio of the amplitude is $\triangle V^2 /8 R k T \beta$ for the oscilloscope and $\triangle V^2 /R k T \sqrt{8 \beta \beta_1}$ for the slow output meter. If $\beta V_0 = 0$ and $\triangle V^2 >> 4 R k T \beta$, these ratio's become $\triangle V_1^! \sqrt{32 R k T \beta}$ and $\triangle V/\sqrt{32 R_0 k T \beta_1}$ respectively. If there is a large unbalance $b V_0 >> \triangle V$ and also if $b^2 V_0^2 >> 4 R k T \beta$, we find

$$\left(\frac{\text{signal}}{\text{noise}}\right)_{\text{c.r.o.}} = \frac{\triangle V}{\sqrt{8 R_0 k T \beta}}$$

$$\left(\frac{\text{signal}}{\text{noise}}\right)_{\text{meter M}} = \frac{\triangle V}{\sqrt{8 R_0 k T \beta_1}} \qquad (3.15)$$

If the meter M has the characteristic pass band of a critically damped galvanometer, with a period of the free system equal to τ, (3. 15) goes over into

$$\left(\frac{\text{signal}}{\text{noise}}\right)_{\text{meter M}} = \frac{\triangle V \sqrt{\tau}}{\sqrt{2 \pi R_0 k T}} \qquad (3.16)$$

We see that the cases with a large unbalance are the most favorable. If the observation is made with the phase sensitive audio amplifier and τ is the time necessary to take a reading, equal to the RC-value of the filter, we find for the case of large unbalance

$$\left(\frac{\text{signal}}{\text{noise}}\right)_{\text{meter L}} = \frac{V_0 m b \sqrt{\tau}}{\sqrt{4\pi R_0 k T}} \qquad (3.17)$$

We now substitute the expression for the signal $\triangle V$ (3. 4) into (3. 16). A limit is set to the maximum value of the occurring products $u = \chi' V_0$ and $v = \chi'' V_0$ by the relaxation time T_1. The voltage across the coil is, of course, proportional to the amplitude of the radiofrequency field H_1. If we assume the field H_1 to be uniform over the whole volume V_c of the coil, we find

$$V_0 = H_1 \, \omega_0 \, \sqrt{V_c \, L / 4 \, \pi} \qquad (3.18)$$

In chapter 2 we already found that the maximum value for $\chi' H_1$ or $\chi'' H_1$ is $\frac{1}{2} \chi_0 H_0 T_2^{1/2} T_1^{-1/2}$. Inserting this value and (3.18) into (3.16) we find after elimination of ω_0 and L

$$\frac{\text{signal}}{\text{noise}} = \frac{1}{6} \, q \, N \, Q^{1/2} \, V_c^{1/2} \, \gamma^{5/2} \, H_0^{3/2} \, T_2^{1/2} \, T_1^{-1/2} \, \tau^{1/2} \, \hbar^2 I(I+1)(k\,T)^{-3/2} \, F^{-1/2}$$
$$(3.19)$$

Using the modulation method we have to insert (3.11) and (3.18) into (3.17) and then to determine the maximum value of

$$\frac{\partial \, \chi''}{\partial \, H_0} H_s H_1$$

We find $\left(\dfrac{\partial \, \chi''}{\partial \, H_0} H_s H_1\right)_{max} = \frac{1}{4} \chi_0 H_0 T_1^{-1/2} T_2^{1/2} (\gamma \, T_2 \, H_s)$ \qquad (3.20)

The sweep amplitude H_s cannot be chosen larger than the line width $2/\gamma \, T_2$. If we take this value, we already should take into account higher order terms in the expansion (3.9). So we introduce a number λ, smaller than one, and put $\gamma \, T_2 \, H_s = 2 \, \lambda$. We obtain for the signal to noise ratio with the modulation method

$$\frac{\text{signal}}{\text{noise}} = \frac{1}{12} \sqrt{2} \, \lambda \, q \, N \, Q^{1/2} \, V_c^{1/2} \, \gamma^{5/2} \, H_0^{3/2} \, T_2^{1/2} \, T_1^{-1/2} \, \tau^{1/2} \, \hbar^2 I(I+1)(kT)^{-3/2} \, F^{-1/2}$$
$$(3.21)$$

If we use the bridge of fig. 3.4 instead of fig. 3.3 we have to add a factor $2^{-1/2}$ to (3.19) and (3.21).

Furthermore we have already added a factor $F^{-1/2}$; F is a number larger than one and is called the noise figure of the receiver. Due to the shot effect of the current in the tubes and the Brownian motion in the resistors, the receiver itself is also a source of noise. The over-all effect of this additional noise on the precision is described by the quantity F, defined by

$$\frac{\text{signal output power of generator}}{\text{noise output power of generator}} = F \, \frac{\text{signal output power of receiver}}{\text{noise output power of receiver}}$$

or F is the number, by which the absolute temperature of the internal resistance of the generator has to be multiplied, if we want to ascribe all the noise to that resistance. The factor $(kT)^{-3/2} F^{-1/2}$ in (3.19) and (3.21) must be split in two parts.

In $(kT)^{-1}$ the temperature T refers to the sample, and in $(kFT)^{-1/2}$ the effective noise temperature of the bridge is denoted by FT.

The noise figure is not a constant of the receiver, but depends on the impedance across the receiver input, and may also be a function of the frequency. It can be measured by the circuit of fig. 3.12. The tube is a

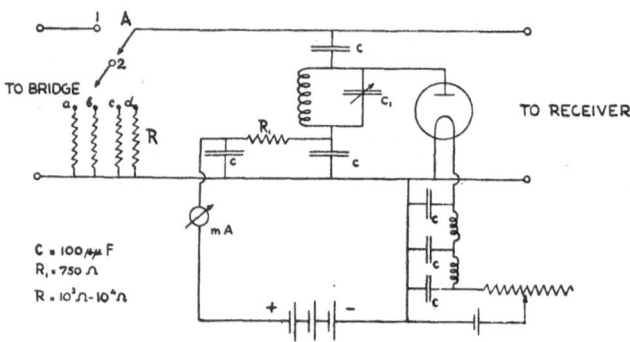

Figure 3.12.

The noise diode. The filament is heated by a variable current, which passes through an r.f.filter. The plate current also passes through an r.f.filter. The impedance of the LC circuit is high compared to the various resistors R, parallel to it. With the switch A we can also put the bridge impedance in parallel to the tuned plate impedance. Further explanation is given in the text.

diode with tungsten filament, the temperature of which can be varied. The diode current i_d, measured by the ma-meter, is always temperature limited. We switch A into position 2 and measure the diode current necessary to double the noise power output of the receiver for various values of R. From this we obtain the noise figure as a function of R according to

$$\frac{2\,e\,i_d\,R^2\,\Delta\nu}{4\,R\,k\,T\,\Delta\nu} = \frac{4\,R\,k\,(FT)\,\Delta\nu}{4\,R\,k\,T\,\Delta\nu} = F$$

Usually we find that the function $F(R)$ has a minimum. Then we switch A into position 1 and measure the noise figure with the bridge impedance. If the value of F so found is not close to the minimum value, the impedance of the bridge should be transformed, e.g. by another choice of the coupling condensers. The condenser C_1 is in every case adjusted for resonance. The noise figure for the pre-amplifier in combination with the bridge was found to be 2, for the National HRO 5 receiver about 12.

We now summarize the results of this section. The precision of the method is limited by the general thermodynamical fluctuations in voltage across a resistance. These could be reduced in our case by lowering the temperature of the bridge impedance. At the same time, however, care should be taken that the relative influence of noise sources in the receiver, expressed by the noise figure F, remains low. Formulae for the signal to noise ratio have been derived. It appears from (3.16) that the radio-frequency bridge must never be completely balanced; moreover, a stable complete balance would be hard to attain experimentally. With sufficient unbalance the signal to noise ratio is independent of the degree of balance b. Then the nuclear signal is detected by mixing with the main carrier rather than by a square law detector.

By observation on the audio output meter L the ratio is independent of the band width of the receiver. Increasing the precision by narrowing the pass band of the audio amplifier, is equivalent with an increase in the time required to make one measurement. The main advantage of the phase sensitive audio amplifier is not a better signal to noise ratio, for according to (3.19) and (3.21) one looses a factor $\sqrt{2}\lambda$ compared to the reading on meter M in the detector circuit, having the same time of indication. But the apparatus becomes less sensitive to external disturbances, as these usually do not have $30 \sim$ components. The effects of drifts in generator output and detector current are eliminated to a great extent. The zero reading is steadier. The balance of the bridge is necessary, as it is hard to find a detector, which can indicate changes of one part in 10^5 and still has a good noise figure. The balance reduces further the influence of drifts and frequency modulation in the output of the signal generator. From (3.21) we see that for the same coil the accuracy increases as $\nu_0{}^{7/4}$, as H_o is proportional to the resonance frequency ν_0 and Q to $\nu_0{}^{1/2}$. Since Q is also proportional to the linear dimensions of the coil, the accuracy increases proportional to $V_c{}^{2/3}$, when the coil is enlarged.

For large coils, however, more current is needed to produce the same field density, and therefore higher demands are put on the balance. Also the field H_o has to be homogeneous over a larger area. Finally we want to stress the importance of having a good filling factor q and a high density of nuclei N.

3. 4. *Measurement of the line width and relaxation time.*

In this section we assume that the magnetic field H_o is perfectly homo-geneous. Suppose that we have adjusted the frequency of the signal generator and the magnetic field H_o approximately to the gyromagnetic ratio of the nuclei in the sample in the coil. On applying a modulation of sufficient amplitude to the field we shall pass through the resonance twice in each cycle. On page 71 an oscillogram of a resonance line at 29 Mc/sec of protons in 0.4 g glycerin is represented. Figure 3. 13 shows the absorption, figure 3. 14 the dispersion. The pictures were taken with a linear time base, while the field modulation was a 30 $\sim$ sine function. If we also put a 30 $\sim$ sine sweep on the horizontally deflecting plates of the oscilloscope in phase with the other sweep, we obtain a linear scale in oersted. The oscillogram in fig. 3. 15 shows a resonance line in the forth and back sweep. The curves do not coincide because a small phase shift was left on purpose. The peculiar wiggles will be discussed at the end of this chapter. We determine the amplitude of the sweep by using a pick up coil of 1200 turns with an average area of 3 cm². The 30 $\sim$ voltage induced in this coil, when it is put in the magnet gap is measured with a Ballantine vacuum tube voltmeter. Another way to calibrate the horizontal scale on the oscilloscope in oersted is to change the radio frequency, say, by 0.05 %. After rebalancing the bridge the resonance line appears on the screen somewhat shifted with respect to its original position. The displacement corresponds to 0.05 % of the total magnetic field.

Thus we are able to express the distance between the points of half the maximum value in oersted. The relation between this ΔH and T_2 for a damped oscillator curve is (compare section 2. 6),

$$\gamma \triangle H = 2/T_2$$

and for a Gaussian

$$\gamma \triangle H = 4 \, ln \, 2/T_2$$

(3. 22)

With the audio amplifier the line width can be measured too. A sweep of a fraction of the line width is used and the total magnet current, deter-mining H_o, is changed in small steps. The calibration can again be made by the shift of the resonance with a small variation in the frequency. For variations of the magnetic field of less than 0.05 % a linear relation between the current and the field was found. The determination of the line width in this way was less accurate, as slight hysteresis effects can not be entirely excluded. It was only applied to wide and so usually very weak lines, which were hardly visible on the oscilloscope because of the

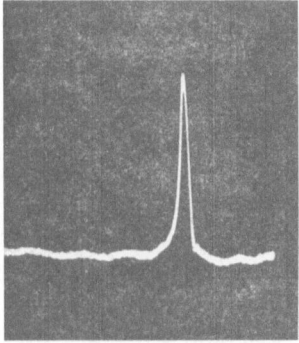

Figure 3. 13.

Oscillogram of the nuclear magnetic resonance absorption of protons in glycerin at 29 Mc/sec.

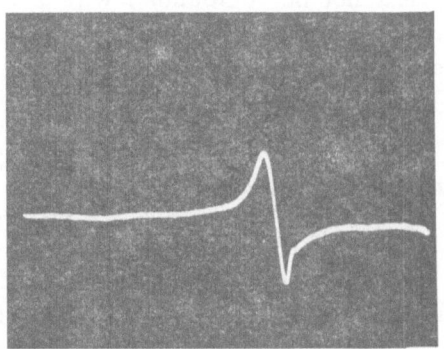

Figure 3. 14.

Oscillogram of the nuclear magnetic resonance dispersion of protons in glycerin at 29 Mc/sec.

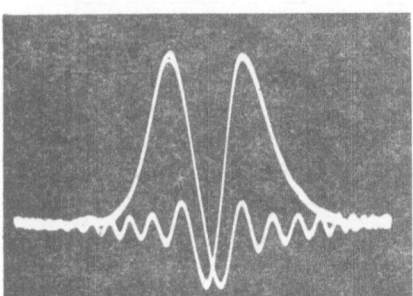

Figure 3. 15.

Oscillogram of the nuclear magnetic resonance absorption of protons in water at 29 Mc/sec. The peculiar wiggles are a transient effect, which is discussed in section 3. 7. Two resonances occur because a sinusoidal sweep was used with a small phase shift, so that the curves do not coincide. The wiggles always appear after the sweep has passed through resonance.

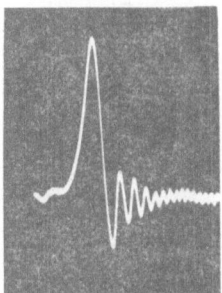

Figure 3. 16.

Oscillogram of the nuclear magnetic resonance absorption of protons in water at 29 Mc/sec. The wiggles are shown for a linear sweep.

noise background. In fig. 3. 17 a typical derivative absorption curve is plotted for the proton resonance in a 0.5 N solution of Fe $(NO_3)_3$. It must not be confounded with a dispersion curve! The distance $\Delta H'$ between the maximum and minimum in these curves, that is between the points of maximum slope in the original absorption curve, is related with T_2 for the damped oscillator by

$$\gamma \triangle H' = 2/\sqrt{3}\, T_2$$

and for the Gaussian by $\qquad\qquad\qquad\qquad\qquad\qquad\qquad$ (3. 23)

$$\gamma \triangle H' = 2/T_2$$

The actual shape of the resonance curves is discussed in detail by Pake (P 1). It cannot be easily distinguished from the experimental curves, which type of curve we have. Dispersion curves would give a better criterium. A Gaussian distribution is the most likely.

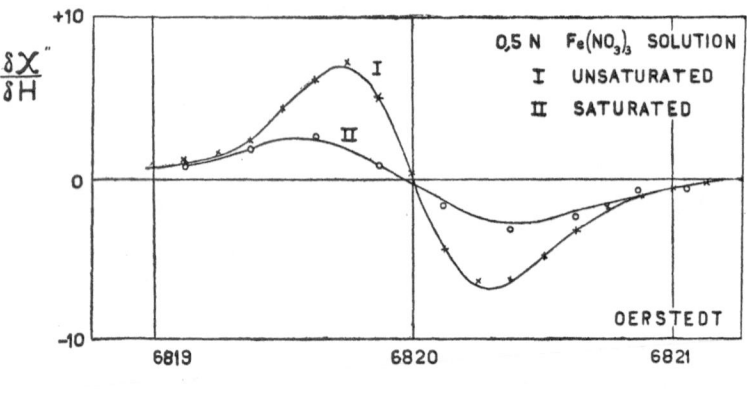

MAGNETIC FIELD H_o

Figure 3. 17.

The derivative of the magnetic resonance absorption of protons in a Fe $(NO_3)_3$ solution, measured with the phase sensitive audiomixer.

If the width is measured in stronger radio frequency fields a broadening of the line by saturation will set in according to (2. 63). The saturation effect enables us to measure the relaxation time T_1. The absorption curve is measured at various values of the output of the signal generator. Off resonance the detector current meter M reads proportional to this power, that is proportional to V_0^2. However, it the power of the generator has been turned up, we turn the gain of the receiver down, so that M reads its original value. In this way a direct influence of the change in the generator output on the reading of the meters M and L is eliminated. The modulation H_s and the balance of the bridge are kept unchanged. From (3.8) we see that the deflection of the audio meter L will be proportional to the 30 $\sim$ modulation, and under the conditions mentioned will be proportional only to $\dfrac{\partial \chi''}{\partial H}$.

Usually we do not make use of the whole curve, but merely determine the maximum deflection in either direction. A simple calculation *) shows that the extreme values in the derivative of the absorption of the damped oscillator type (2.63) decrease with increasing H_1 yielding a deflection proportional to

$$(1 + \gamma^2 H_1{}^2 T_1 T_2)^{-3/2} \tag{3.24}$$

The extreme in the derivative of the dispersion should decrease only as $(1 + \gamma^2 H_1{}^2 T_1 T_2)^{-1}$. So for higher degrees of saturation it becomes more and more important to have pure balance, as the effect of χ' decreases more slowly than of χ''. For the Gaussian distribution it is not possible to give the decrease in the meter reading in a closed form, but the general behaviour is the same. The reading starts to decrease rapidly with increasing power, when $\gamma^2 H_1{}^2 T_1 T_2$ becomes of the order of unity. In any case, for the same line shapes $\dfrac{\partial \chi''}{\partial H}$ will be the same function of $\gamma^2 H_1{}^2 T_1 T_2$ and H_1 is proportional to the output voltage of the generator. If we plot therefore the maximum deflection of meter L against the reading of the output meter of the signal generator on a semi logarithmic scale, we obtain a series of parallel curves, which decrease rather suddenly around the value of H_1 given by $\gamma^2 H_1{}^2 T_1 T_2 = 1$. In fig. 3.18 some typical curves are drawn, obtained for saturation of protons in ice at various temperatures. For comparison the theoretical curve $(1 + x^2)^{-3/2}$ is indicated. Now twice the horizontal distance between the curves for different samples will give us the ratio of the products $T_1 T_2$ in these samples. And if T_2 has been measured in an independent way, we obtain the ratio of the relaxation times T_1. An absolute determination would be possible if the value of H_1 in the coil were known. Of course we know its order of magnitude from the circuit constants, but it is hard to make an estimate of the insertion and coupling losses. Therefore it was decided to make an absolute determination of the relaxation time in one substance along other lines.

---

*) Strictly speaking, formula (3.24) holds only for the case $\omega_s T_1 \ll 1$. For $\omega_s T_1 \gg 1$ the saturation is described by a different expression, which falls off somewhat slower than (3.24) with increasing H_1.

Distilled water was taken, since it was known that its relaxation time is of the order of a second, and so long enough to apply the following method.

On the screen of the oscilloscope the proton resonance was observed with a large amplitude of the sweep (about 5 oersted). The power of the generator was chosen such that no saturation occurred but was about to set in. Then the sweep amplitude was turned down to 0.5 oersted, but still the whole line, which was narrower than that value, was covered. Since the time spent at resonance was then 10 times longer, saturation would occur. After that the sweep amplitude was suddenly turned up to its original value. The height of the absorption peak immediately after this would be small, since the surplus number of protons was reduced by the preceding saturation. Exponentially the absorption peak would increase to its unsaturated value.

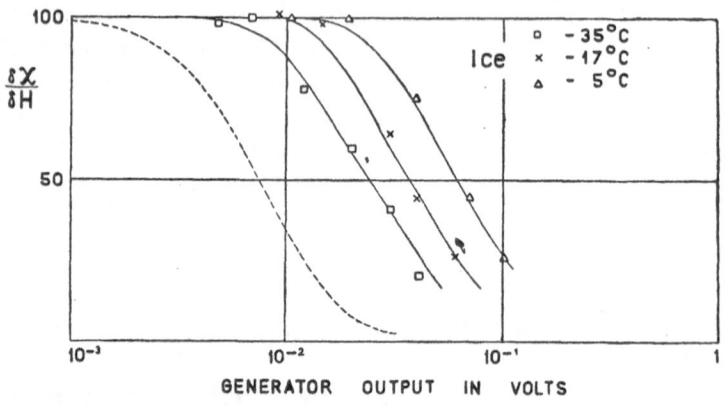

Figure 3. 18.

The saturation effect of the proton spin system in ice. The maximum of the derivative of χ'' is plotted, essentially against the amplitude of the alternating field H_1. The dotted curve, representing $(1 + x^2)^{-3/2}$, is indicated for comparison with theory. Data like these, in combination with data on the line width, were used to construct fig. 4. 9.

This last process was filmed with a movie camera and yielded 2.3 ± 0.5 sec. for the relaxation time in water. The same experiment was done for petroleum ether, giving 3.0 sec, and for a 0.002 N solution of $Cu\,SO_4$ in water we found 0.75 sec. All other relaxation times were measured relative to these values.

3. 5. *The inhomogeneity of the magnetic field.*

It turned out that in many substances, especially in liquids and gases, the measured line width was caused by the inhomogeneity of the field. The coil with sample was moved around in the magnet gap until the narrowest line was obtained. The best spot in our magnet appeared to be closer to the edge than to the centre, and there is reason to believe that the inhomogeneity is not so much caused by the geometrical conditions, as by inhomogeneities in the iron pole pieces. The best condition obtained was an inhomogeneity over the region of the sample of 0.12 oersted in a total field of 7000 oersted, and an inhomogeneity of 0.015 oersted in a field of 1100 oersted. This latter field was used for proton resonances at 4.8 Mc/sec. All lines taken in this spot in the gap, which are wider than these limits, show of course their real width. There are reasons to believe that the line width, e.g. of water, is still much narrower than 0.015 oersted.

◄2 OERSTEDT►

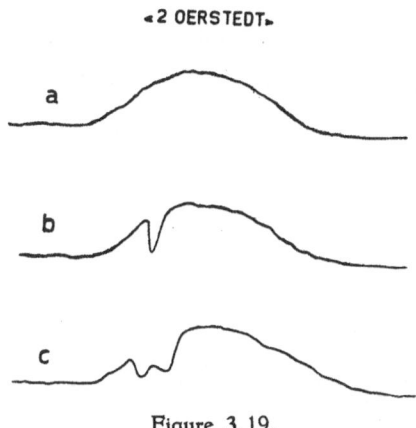

Figure 3. 19.

The absorption of protons in water in an inhomogeneous field, as observed on the screen of an oscilloscope.
a. Unsaturated.
b. Partial saturation. The dip disappears in about the relaxation time.
c. Partial saturation over a larger region. Where the residual sweep went slow and reversed, the saturation is more complete.

The experimental results and the theory of the line width will be further discussed in the following chapter. Here we must consider, what happens to the saturation experiment, when we are forced to measure the resonance in an inhomogeneous field which determines the line shape. First we shall describe an experiment, that was of „historical interest" in the discovery of the narrow resonance lines. Suppose we have a water sample in a rather inhomogeneous field, so that the line width is two oersted. The sweep amplitude is originally turned off and the system of those nuclei, which are at resonance, is saturated. If one then suddenly turns on a sweep of 5 oersted, one sees a "hole" in the resonance line, which disappears in about 2.3 sec, the relaxation time (fig. 3. 19 b). One can also leave a small residual sweep and saturate the spin system over a somewhat broader range of the inhomogeneous field (fig. 3. 19c). The little dips indicate more saturation at the points where the sweep goes slower and reverses, that is where the nuclei are longer at resonance.

For the measurement of the relaxation time by the decrease of $\left(\dfrac{\partial \chi''}{\partial H}\right)_{max}$ we must distinguish three cases.

a. $\triangle h \ll H_s \ll 1/\gamma T_2$. Here $\triangle h$ is a measure for the inhomogeneity of the field which is described by the distribution function $h(\nu)$. In this case we can consider $h(\nu)$ as a δ-function and the considerations of the preceding section with formula (3.24) are valid.

b. $H_s \ll 1/\gamma T_2 \ll \triangle h$. As here the total sweep is still less than the natural line width, we can integrate the result for a single line derived in the preceding section over the distribution in the field. The reading v will be proportional to

$$v \sim \int_0^\infty \frac{\partial \chi''}{\partial H} h(\nu') \, d\nu'$$

Remembering that $H = 2\pi\nu/\gamma$ and $\dfrac{\partial}{\partial \nu} = -\dfrac{\partial}{\partial \nu'}$ we obtain by partial integration for the deflection

$$v \sim \int_0^\infty \chi'' \frac{\partial h}{\partial \nu'} \, d\nu'$$

If $\dfrac{\partial h}{\partial \nu'}$ changes slowly over the region of the natural width we find:

$$v \sim h'(\nu)(1 + \gamma^2 H_1^2 T_1 T_2)^{-1/2} \tag{3.25}$$

So the signal should decrease more slowly with increasing H_1 than in case a.

c. $1/\gamma T_2 \ll H_s \ll \triangle h$.

In this case the amplitude of the sweep is large compared to the natural width. We assume for the moment, that the sweep comes back before the nuclear system has had a chance to relax $\omega_s \gg 1/T_1$. We can define an average field density as the product of the actual density and the fraction of the time, spent at resonance. We have something as is indicated in fig. 3.19c. We are going back and forth in the bottom of the hole and have to determine the 30 $\sim$ component. For the sake of simplicity let us assume that the sweep has a constant velocity given by $\dfrac{\partial H}{\partial t} = H_s \omega_s$ and that the line shape is a rectangle with width $2/\gamma T_2$. The fraction of the time spent at resonance is then given by $1/2\pi \gamma T_2 H_s$.

Repeating an argument similar to that leading to (2.70) we find that the surplus number of nuclei over the region of the sweep is reduced by a factor $1/(1 + \gamma^2 H_1^2 T_1/2 \pi \gamma H_s)$ and the deflection of the meter is proportional to

$$v \sim \frac{h'(\nu)}{1 + \gamma^2 H_1^2 T_1/2 \pi \gamma H_s} \qquad (3.26)$$

It has been observed experimentally that the value of H_1, necessary to produce saturation, depends on the amplitude of the sweep as must be expected from (3.26). If H_s is kept constant for various samples, we obtain a set of parallel saturation curves, which should decrease as $(1 + x^2)^{-1}$.

If we plot the curves again on semi logarithmic paper, twice the horizontal distance between them gives immediately the ratio of the relaxation times T_1. If H_o is varied in order to go through the curve $h(\nu)$ in fig 3.19, we meet unsaturated groups of nuclei and we must "dig a hole" in the distribution $h(\nu)$, before we obtain the equilibrium value *).

So far we have assumed that $H_o + H_s \sin \omega_s t$ is slowly varying, so that we could make use of the stationary solution of Bloch's equations. For narrow lines this assumption is not justified. The modulation of the field should be taken into acount by considering a frequency-modulated spectrum of the radio-frequency signal.

To estimate the order of magnitude of the transient effects, which will be discussed somewhat further in section 3.7, we assume that we have only one passage of the sweep with a velocity $H_s \omega_s$. The time spent at resonance is not, as was suggested, $1/2 \pi \gamma H_s \omega_s T_2$. For if the nuclei experience a signal only during a time t', the angular frequency is not defined better than to $\triangle \omega \approx 1/t'$, and if $1/T_2 < \triangle \omega$ we should write $1/t' = \triangle \omega = \gamma H_s \omega_s/\triangle \omega$. Substituting the experimental values $H_s = 0.02$ oersted, $\omega_s = 60 \pi$ sec^{-1} we find for protons $\triangle \omega = 550$ sec^{-1}. Since $\gamma H_s = 600$ sec^{-1}, the nuclei are at resonance practically during the

---

*) This „hole" effect makes it necessary to be very careful in carrying out the experiment.

If we go through the resonance too fast, in a time comparable to the relaxation time, we are always operating on the right hand side of the hole with increasing H_o and on the left hand side with decreasing H_o. So we get large readings on meter L in opposite directions depending on the sign of the variation in H_o. So we have to go very slowly, but we cannot go too slow either, because shifts in the battery current will alter the magnetic field by a few parts in a million over periods of about half a minute and so cause jumps to other places in the line.

whole sweep. The effective energy density of the field is $\varrho\,(\nu) = \dfrac{H_1^2}{8\,\pi}\,\dfrac{2\,\pi}{\triangle\,\omega}$
and the saturation should be described by a function

$$v \approx \frac{h'\,(\nu)}{1 + \pi\,\gamma^2\,H_1^2\,T_1/\sqrt{\gamma H_s}\,\omega_s} \qquad (3.27)$$

which displays the same dependence on H_1 and T_1 as (3.26). We now got rid of the restriction $\omega_s \gg 1/T_1$ for $T_2 > 1/550$ sec. If this condition is not satisfied, we immediately come back to the cases a or b. The intermediate case b should occur for values of $3.10^{-4} < T_2 < 2.10^{-3}$ but experimentally it could not be distinguished from case a, which occurs for $T_2 < 3.10^{-4}$ sec.

The ratios of the relaxation times T_1 were determined with formula (3.24) for $1/\gamma\,T_2 > H_s$ and with (3.26) for $1/\gamma\,T_2 < H_s$

This method gave very satisfactory results which are shown in the graphs of chapter 4. But we cannot exclude with certainty a serious systematic error in the determination of relative relaxation times in the region of transition, for intermediate values of T_2. The error, however, is probably less than a factor 2. For in the preceding discussion we have made some drastic simplifications, that the line shape is rectangular and the sweep linear. In fact the sweep is sinusoidal and we do not have a single passing through the resonance, so that the approximation of a signal smeared out over a frequency range $\triangle\omega$ is rough. Neither did we discuss the transitions from one case to another. Furthermore we have assumed that the oscillating field H_1 has a constant amplitude over the region of the sample. At the ends of the coil, however, H_1 will certainly be smaller. For broad lines, where H_o can be considered as homogeneous, this will tend to make the saturation curves less steep, since the systems of nuclei in various parts of the sample are not saturated at the same current in the coil. If both H_o and H_1 are inhomogeneous, the situation is very complicated. Suppose that in the middle of the coil H_o has its highest value, at the ends its lowest. Then the inhomogeneity of H_1 will even influence the shape of the unsaturated line, and on saturation we cannot expect to measure a curve given by e.g. (3.25).

Experimentally we found that the saturation curves for small values of $1/T_2$ (case c) were almost parallel to those for lines with real width in case a. The small deviation would be to the other side than predicted by (3.24) and (3.27). The curves would be steeper than $(1 + x^2)^{-3/2}$. In fig. 3.20 we give a typical example. The relaxation times, obtained from these curves and many others, will be described and compared with theory in the next chapter. In fig. 3. 21 the roughest set of curves which has been used in the evaluation of the data is represented, so that the reader may form his own opinion on the obtained accuracy.

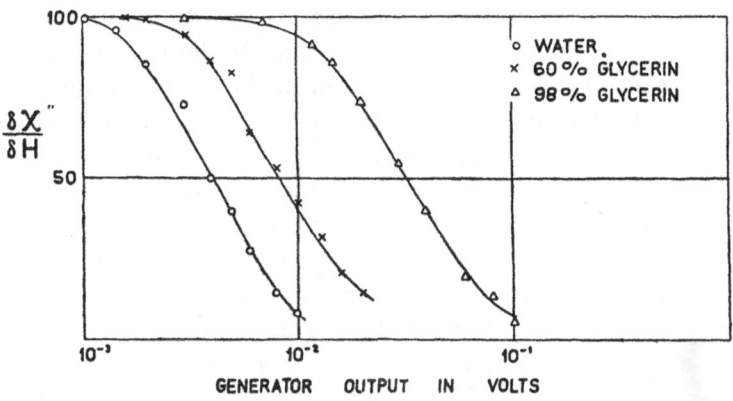

Figure 3. 20.

Saturation of the magnetic resonance absorption of protons in mixtures of water and glycerin.

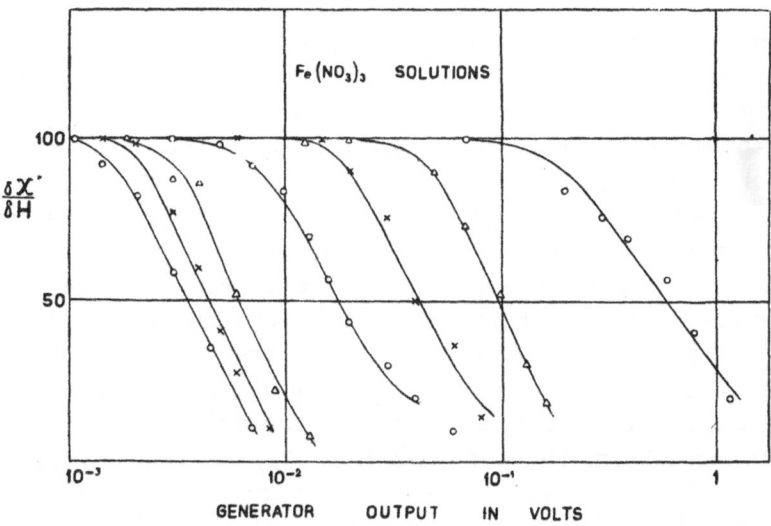

Figure 3. 21.

Saturation of the magnetic resonance absorption of protons in solutions of Fe (NO₃)₃ of various concentrations. Data like these were used to construct fig. 4. 5.

Any systematic error could be eliminated if the field were made so homogeneous, that always the real width would be measured. In the relative measurement within the group of lines wider than 0.3 oersted, we do not expect systematic deviations. These are also eliminated in comparing the lines, which are narrower than 0.01 oersted. In these cases the error arises from the inaccuracy in the calibration of the radio frequency power meter and in the decade system to reduce the output of the signal generator. According to General Radio Co. this systematic error can amount to about 20 % in power at 30 Mc/sec. Slight drifts in balance of the bridge and gain of the receiver in the time necessary to plot a curve might affect the result by 15 %. For wide lines another 10 % has to be added for inaccuracy in the determination of T_2. In the measurement of the line width we can also expect systematic errors, caused by the line shape or the finite velocity of the sweep. The noise was always less than 10 % and in most experiments less than 1 % of the reading. Since each relaxation time is determined from an entire saturation curve, accidental errors will average out to some extent. Several runs were repeated on different days. The resulting relaxation times were reproducible within 30 %.

3. 6. Comparison with the "nuclear induction" experiment.

There has been some confusion about the question, how the method of P u r c e l l, T o r r e y and P o u n d (P 7) compares to that of B l o c h, H a n s e n and P a c k a r d (B 7), which is called the method of nuclear induction by those authors. It may be pointed out here that there is no essential difference. B l o c h c.s. pick up the nuclear signal in a separate coil, the axis of which is perpendicular to H_o and to H_1. P u r c e l l uses the same radiofrequency coil for supplying the field and picking up the signal. We have seen in chapter 2 that the nuclear resonance depends only on one rotating component of H_1. Therefore the signal picked up in any coil perpendicular to H_o will be the same. Nothing prevents us from supposing a second coil to be present in the P u r c e l l experiment, just as in B l o c h's arrangement. In this coil are flowing two equal and opposite currents, providing a field $H_y = \frac{1}{2} H_1 \sin \omega t$ and a field $H_y = -\frac{1}{2} H_1 \sin \omega t$. Since the sum of the currents is zero, we can leave the coil out. The only advantage to pick up the signal in a separate perpendicular coil is that one has automatically achieved a balance by the geometrical arrangement. For very high values of H_1 (> 1 oersted), where B l o c h, H a n s e n and P a c k a r d carried out their experiments, this is probably the only way of attaining the required balance. They also operate, however, for the same reason as we do, with some residual unbalance. On the type of unbalance it will depend again, whether the real or imaginary part of the susceptibility is measured. But at very high values of H_1 the absorption goes to zero, so that one can only measure the dispersion anyhow (G 5, B 7). At these high field strenghts information about the natural line width and relaxation time can then only be obtained by making use of nonstationary conditions, in which the frequency and amplitude of the sweep are varied. Once more it may be asserted that exactly the same phenomena would occur in the Purcell experiment with the same values of the parameters.

3. 7. Transient effects.

The non-stationary conditions, caused by the finite speed of the sweep and which are made use of in Bloch's method, cause some undesirable transient effect in our arrangement. If the time spent at resonance in the course of the sweep becomes short compared to T_1 and (or) T_2, it is not permissible to solve Bloch's equations (2. 59) under the assumption, that H_z is independent of time. We shall not go into the theory, but must point out that the picture of the observed proton resonance in water (figs 3.15 and 3.16) can be explained by these non-stationary conditions. The wiggles occur after having passed through resonance. At resonance all nuclei are brought in phase by the applied radio-frequency signal. When the sweep goes on, the Larmor frequency of the nuclear precession changes and the nuclear signal will beat with the applied signal, causing alternating minima and maxima. The phase angle between the beating signal after passing the resonance at $t = 0$ is given by

$$\varphi = \int_0^t \gamma \left\{ H\left(t\right) - H_0 \right\} d\,t$$

For a linear sweep φ is a quadratic function of the time. The time lapse between successive wiggles varies in the predicted way as a function of the amplitude or frequency of the sweep. The amplitude of the wiggles decays, as the nuclei get out of phase with one another. The decay time is T_2 or in an inhomogeneous field $1/\gamma \Delta h$, whichever is the shortest. The decay is partly caused by the narrow pass band of the receiver, as was checked by the use of a receiver with a broad band of 100 kc/sec. It was even possible to tune the narrow band receiver to the Larmor frequency in the wiggles, so that they become more pronounced relative to the main absorption line. If $H\left(t\right)$ is an increasing function, the wiggles, always occurring after the resonance, arise from signals of higher frequency,

6

with decreasing $H(t)$ from signals with lower frequency than the applied signal. To the observed width of the main absorption line on the oscilloscope is also set a lower limit by the time spent at resonance, as we discussed already in section 3.5. If this time is 10^{-2} sec., the measured width cannot be narrower than 100 cycles/sec. or about 0.02 oersted for protons. With the sweep $H_s = 0.02$ oersted, $\omega_s = 60\,\pi$ sec^{-1}, used in the relaxation experiments in very narrow lines, the signals get hardly out of phase before the sweep turns back, so that the measurements of T_1 are not much affected by the wiggles.

The wiggles disappear for very small amplitude or low frequency of the sweep. Furthermore the meter L is sensitive only to the 30 $\sim$ component in the energy absorption and will not be seriously affected by the transient effect of the wiggles.

CHAPTER 4

THEORY AND EXPERIMENTAL RESULTS.

4.1. *Relaxation time and line width in liquids* (B 10, B 11).

4.1.1. *The Fourier Spectrum of a random function.*

In chapter 2 a general theory for the relaxation time was presented. In order to apply it to practical cases we have to evaluate the Fourier spectra of the functions of the position coordinates F_0, F_1 and F_2 of section 2.5.

In a liquid these functions will vary in a random fashion with time, as the particles containing the magnetic nuclei take part in the Brownian motion. The fluctuating functions $F_0(t)$, $F_1(t)$ and $F_2(t)$ satisfy the condition

$$\overline{\text{Re } F(t)} = \overline{\text{Im } F(t)} = 0 \qquad (4.1)$$

The statistical character of the motion justifies an assumption, customary in the theory of fluctuation phenomena, that

$$\overline{F(t) F^*(t + \tau)} = k(\mid \tau \mid) \qquad (4.2)$$

The left hand side is called the correlation function of $F(t)$.

The correlation function of the random function $F(t)$ is independent of t and an even function of τ. From these assumptions it follows immediately that $k(\tau)$ is real. We shall now derive briefly the relation between this correlation function and the intensity of the Fourier spectrum of $F(t)$. A very general theory of random processes has been given by W a n g and U h l e n b e c k (W 2, R 4), where the reader may find further references. Many other investigators have pointed out the connection between the spectrum and the correlation function. We shall here follow closely K e l l e r 's (K 1) argument, although there are some slight modifications, as we want to distinguish between positive and negative frequencies and our function $F(t)$ is complex. Expand $F(t)$ in a Fourier integral

$$F(t) = \int_{-\infty}^{+\infty} A(\nu)\, e^{2\pi i \nu t} d\nu$$

$$A(\nu) = \int_{-\infty}^{+\infty} F^*(t)\, e^{-2\pi i \nu t} dt$$

(4. 3)

We assume that $F(t) = 0$ for $|t| > T$, where T is a time large compared to all times in which we ever have made or shall make observations. This assumption therefore will not alter the physical results, and in the end we can get rid of it by taking the limit $T \to \infty$. Between the functions connected by the transformation of Fourier (4. 3) exists the P a r s e v a l relation

$$\int_{-\infty}^{+\infty} F(t)\, F^*(t)\, dt = \int_{-\infty}^{+\infty} A(\nu)\, A^*(\nu)\, d\nu$$

(4. 4)

With (4. 3) and our assumption we can write this in the form

$$\overline{F(t)\, F^*(t)} = \frac{1}{2\,T} \int_{-\infty}^{+\infty} d\nu \int_{-T}^{+T}\int_{-T}^{+T} F(t)\, F^*(t')\, e^{2\pi i \nu\,(t-t')}\, dt\, dt'$$

(4. 5)

We next make the substitutions $\sigma = t$ and $\tau = t - t'$. Using the fact that $\overline{F(\sigma)\, F^*(\sigma - \tau)}$ is only different from zero for small values of $|\tau|$, at any rate much smaller than T, we obtain after some calculation

$$\overline{F(t)\, F^*(t)} = \int_{-\infty}^{+\infty} d\nu \int_{-2T}^{+2T} e^{2\pi i \nu \tau}\, F(\sigma)\, F^*(\sigma - \tau)\, d\tau$$

$$= \int_{-\infty}^{+\infty} J(\nu)\, d\nu$$

(4. 6)

with the expression for the spectral intensity

$$J(\nu) = \int_{-\infty}^{+\infty} k(\tau)\, e^{2\pi i \nu \tau}\, d\tau$$

(4. 7)

Since $k(\tau)$ is real and even, $J(\nu)$ is real and even. Because we made a distinction between positive and negative frequencies, the intensity in (4. 7) is half the value usually found in the literature. In the following discussion we shall see that $k(\tau)$ often has the form:

$$k\left(\tau\right) = \overline{F\left(t\right) F^{*}\left(t\right)} \exp\left\{-\mid \tau \mid /\tau_{c}\right\} \qquad (4.8)$$

The combination of (4.7) and (4.8) yields

$$J(\nu) = 2 \overline{F\left(t\right) F^{*}\left(t\right)} \frac{\tau_{c}}{1 + 4 \pi^{2} \nu^{2} \tau_{c}^{2}} \qquad (4.9)$$

In general we can say that $k\left(\tau\right)$ is a function which goes rapidly to zero, if $\mid \tau \mid$ exceeds a value τ_{c} which is characteristic for the mechanism of the Brownian motion and is called the correlation time. The general behaviour of the Fourier spectrum is therefore such that the intensity $J(\nu)$ is practically constant for low frequencies and falls off rapidly, when $2 \pi \nu \tau_{c} > 1$. The time average $\overline{F(t) F^{*}(t)}$ can be replaced by the statistical average according to a general theorem from statistical mechanics.

4. 1. 2. *Evaluation of the relaxation time in water.*

We start out with one water molecule, surrounded, say, by carbondisulfide, which contains no nuclear magnetic moments. We assume that the rotational magnetic moments of the molecules are also zero. We want to calculate the relaxation time of one proton due to the presence of the other. The functions F consist each of a single term:

$$F_{1} = \sin \vartheta \cos \vartheta \, e^{i \varphi} \, /b^{3} \qquad\qquad F_{2} = \sin^{2} \vartheta \, e^{2 i \varphi} \, /b^{3}$$

where b is the constant distance between the two protons. The rotation of the molecule in the liquid will change the angle between the magnetic field H_{o} and the radius vector connecting the two protons in a random fashion.

The correlation function of the expressions F can be calculated if we adopt the same simple picture as D e b y e (D 2) did in his famous theory of dielectric absorption and dispersion, namely a rigid sphere of radius a in a medium of viscosity η and absolute temperature T. D e b y e applies to this model E i n s t e i n 's theory (E 1) of the Brownian motion. In the case that no external forces besides the thermal collisions are present, the probability to find a fixed axis of the sphere in the solid angle $\sin \vartheta \, d \vartheta \, d \varphi$ is described by the ordinary diffusion equation

$$- \frac{\partial f\left(\vartheta, \varphi\right)}{\partial t} = D \varDelta f(\vartheta, \varphi) \qquad (4.11)$$

The diffusion constant D is given by the general expression

$$D = k\,T/\beta$$

The damping constant β for the rotation of a sphere in a viscous medium was calculated by Stokes: $\beta = 8\,\pi\,\eta\,a$.

The Laplacian Δ acts only on the angle variables ϑ and φ.

A solution of (4. 11) may be written in a series of spherical harmonics $Y_{l,\,m}$:

$$f = \underset{l,\,m}{\Sigma}\, c_{l,\,m}\, Y_{l,\,m}\,(\vartheta,\,\varphi)\; e^{-\,t\,D\,l\,(l+1)/a^2}$$

At $t = 0$ the sphere is in the position ϑ_0, φ_0 and $f = \delta\,(\vartheta - \vartheta_0)\,.\,\delta\,(\varphi - \varphi_0)$. From this condition we find the coefficients

$$c_{l,\,m} = Y_{l,\,m}^{*}\,(\vartheta_0,\,\varphi_0)\,\Bigg/ \int\limits_{0}^{\pi}\int\limits_{0}^{2\,\pi} |\,Y_{l,\,m}\,(\vartheta,\,\varphi)\,|^2\,\sin\vartheta\,d\,\vartheta\,d\,\varphi$$

In order to find the correlation function $\overline{F\,(0)\,F^{*}\,(t)}$ note that $F_1 = b^{-3}\,Y_{2,\,1}\,(\vartheta,\,\varphi)$ and $F_2 = b^{-3}\,Y_{2,\,2}\,(\vartheta,\,\varphi)$.

We have $b^3\,F_1^{*}\,(t) = \int\limits_{0}^{\pi}\int\limits_{0}^{2\pi} f\,Y_{2,\,1}^{*}\,\sin\vartheta\,d\,\vartheta\,d\,\varphi = c_{2,\,1}^{*}\,e^{-\,6\,D\,t/a^2}$

The average has to be taken over all possible initial positions, i.e. over ϑ_0 and φ_0.

The final result is

$$\overline{F_1\,(0)\,F_1^{*}\,(t)} = b^{-6}\,\overline{Y_{2,\,1}^{*}\,(\vartheta_0,\,\varphi_0)\;Y_{2,\,1}\,(\vartheta_0,\,\varphi_0)}\;e^{-\,\dfrac{6\,D\,t}{a^2}} = \frac{2}{15}\,b^{-6}\,e^{-t/\tau_\mathrm{c}}$$

$$\overline{F_2\,(0)\,F_2^{*}\,(t)} = \frac{8}{15}\,b^{-6}\,e^{-t/\tau_\mathrm{c}} \tag{4. 12}$$

$$\text{with } \tau_\mathrm{c} = 4\,\pi\,\eta\,a^3/3\,k\,T \tag{4. 13}$$

The characteristic time of Debye τ we obtain by carrying out the same procedure for the function $\cos\vartheta = Y_{1,\,0}$

The result is

$$\tau = 4\,\pi\,\eta\,a^{\,3}/k\,T = 3\,\tau_\mathrm{c} \tag{4. 14}$$

In Debye's theory τ is the time in which an assembly of water molecules, originally oriented by an electric field, loses its distribution around a preferred direction by the Brownian motion, after the electric field has been switched off. In our case τ_c is the time, in which a molecule is rotated by the Brownian motion over such an angle that the relative position of the nuclei with respect to the external field and thus the functions F have changed appreciably.

Using (4.9), (4.12) and the general formula (2.53) we find for the relaxation time of a proton in a watermolecule

$$(1/T_1)_{rot} = 0.4 \left\{ \frac{\tau_c}{1 + 4\pi^2 \nu_0^2 \tau_c^2} + \frac{2\tau_c}{1 + 16\pi^2 \nu_0^2 \tau_c^2} \right\} \gamma^4 \hbar^2 I_p (I_p + 1) b^{-6} \tag{4.15}$$

Substituting numerical values $T = 300$, $\eta = 10^{-2}$, $a = 1.5 \times 10^{-8}$, $I_p = \frac{1}{2}$ we find that $\tau_c = 0.35 \times 10^{-11}$ sec, and since $\nu_0 = 3 \times 10^7$ cycles/sec we have $2\pi\tau_c\nu_0 \ll 1$. We see from (4.15) that in this case $1/T_1$ is proportional to τ_c and we can write with (4.15)

$$(1/T_1)_{rot} = 0.9 \, \gamma^4 \hbar^2 b^{-6} \tau_c \tag{4.16}$$

The value of $\tau = 3\tau_c \approx 10^{-11}$ sec is in excellent agreement with experimental data on the dielectric absorption and dispersion in water at microwave frequencies (C 5).

Next we consider the practical case that the neighbours are not CS_2 molecules, but other H_2O molecules. We can estimate the influence of the other protons on the relaxation time in the following way.

Again the Brownian motion is responsible for the Fourier spectrum, but the cause is now rather the relative translational motion of the molecules than a rotation. Let us consider the protons in the other molecules as independent of one another [1]. We ask for $\overline{F(t)\,F(t+\tau)}$ and τ_c for the protons in a spherical shell between r and $r + dr$ around the proton of which we wish to determine the relaxation process. A reasonable value for τ_c is apparently the time it takes for a molecule to travel over a distance r. For in that time the relative position and with it the spin spin interaction has changed appreciably. From the theory of Brownian motion we have the expression for the mean square displacement of a particle

$$\overline{x^2} = 2kT\tau_c/\beta \tag{4.17}$$

---

[1] It would be better to consider the molecules as independent and attribute to them a moment $2\mu_p$ if the spins are parallel, or zero if they are antiparallel, and then apply to these moments the statistical weight of the parallel and antiparallel state. The same answer would be obtained. In the preceding problem of the rotating molecule also ortho- and para- states should have been distinguished. We shall come back to this question at the end of chapter 5.

where β is a damping constant. For a sphere in a viscous medium S t o k e s derived $\beta = 6 \pi \eta a$

If one prefers to use the diffusion constant $D = k T/\beta$, we find for the correlation time

$$(\tau_c)_{transl.} = \overline{x^2}/2D = r^2/12D \qquad (4.18)$$

since r is the relative displacement of two particles in any direction.

To find $\overline{F(t)\ F(t)}$ we have to average the angular functions over the spherical shell and multiply with the number of protons in the shell as we treat them independently. Then we have to integrate over r to include all other molecules, so approximately from $2a$, the distance of closest approach, to infinity. Using again (4.9) and (2.53) we find

$$(1/T_1)_{transl.} = 1.6 \pi N \gamma^4 \hbar^2 I_p (I_p + 1) \int_{2a}^{\infty} \frac{r^2}{r^6} \left\{ \frac{\tau_c}{1 + 4\pi^2 \nu_0^2 \tau_c^2} + \frac{2\tau_c}{1 + 16\pi^2 \nu_0^2 \tau_c^2} \right\} dr$$
$$(4.19)$$

In the integral we can neglect the term with $\nu_0^2 \tau_c^2$ in the denominators, since $2\pi \tau_c \nu_0 \ll 1$ for $r < 10^{-7}$, and the most important contribution to the integral comes from the nearest neighbours. Integration of (4.19) then simply leads to

$$(1/T_1)_{transl.} = 0.9 \pi^2 \gamma^4 \hbar^2 \eta N/k T \qquad (4.20)$$

Substituting numerical values in (4.16) and (4.20), $a = 2 \times 10^{-8}$, $b = 1.5 \times 10^{-8}$, $\eta = 10^{-2}$, $N = 7 \times 10^{22}$, $\gamma = 2.7 \times 10^4$ we find

$$(T_1)_{rot} = 5.2 \text{ sec.} \qquad (T_1)_{transl} = 10 \text{ sec.} \qquad T_1 = 3.4 \text{ sec.}$$

This value is in good agreement with the experimental value of 2.3 sec. In the case of a rotating sphere it was possible to calculate the correlation function explicitly. For the translational effect and the rotation of more complicated molecules in liquids this would be very difficult. In these cases one might assume formula (4.8) or a linear combination of them with various τ_c. The correlation time τ_c should be larger in more viscous media as the molecular motion becomes slower. In the next section we shall discuss the general relation between the relaxation and correlation times and the viscosity.

4. 1. 3. *The relation between the relaxation time, the viscosity, the correlation time and the Debye time.*

There may be some doubt whether is is permissible to extend the macroscopic notions of viscosity and diffusion to regions which contain only a few atoms. The same objection can be raised against D e b y e's theory. There, as in our case, the procedure is justified by its success. Since we obtain the right order of magnitude for the relaxation time, we might even inversely use the latter to extend our information regarding the motion of the molecules. From our general considerations we would· expect that the relaxation time would decrease with increasing viscosity, as long as the condition $2 \pi \nu_0 \tau_c \ll 1$ is satisfied. This is confirmed by the experimental evidence in Table I and Table II.

Table I
Relaxation time of protons at 29 Mc/sec in hydrocarbons at 20° C

	Viscosity in centipoises	Relaxation time in seconds
Petroleumether	0.48	3.5
Ligroin	0.79	1.7
Kerosin	1.55	0.7
Light machine oil	42	0.075
Heavy machine oil	260	0.013
Mineral oil	240	0.007

Table II
Relaxation time of protons at 29 Mc/sec in polar liquids at 20° C

	Viscosity in centipoises	Relaxation time in seconds
Diethylether	0.25	3.8
Water	1.02	2.3
Ethylalcohol	1.2	2.2
Acetic acid	1.2	2.4
Sulfuric acid	25	0.7^5
Glycerin	1000	0.02^3

The viscosities in table I were measured with a viscosimeter, (time of flow measurement), those of table II were taken from the Physikalisch Chemische Tabelle.

We also measured the relaxation time in mixtures of water and glycerin, of which the result is shown in fig. 4. 1.

The dependence of the relaxation time on the viscosity is not quite the inverse proportionality, which one might infer from (4. 16) and (4. 20). The relaxation time in glycerin is only 10^2 times smaller than in water, while the viscosity is 10^3 times larger. In the first place one can remark that in going from one substance to another the quantities a, b and N change too. The deviation in sulfuric acid can so partly be understood because the proton density in it is much smaller than in the other substances. But for the latter the density of nuclei nor the internuclear distances b change very much from molecule to molecule. The molecular

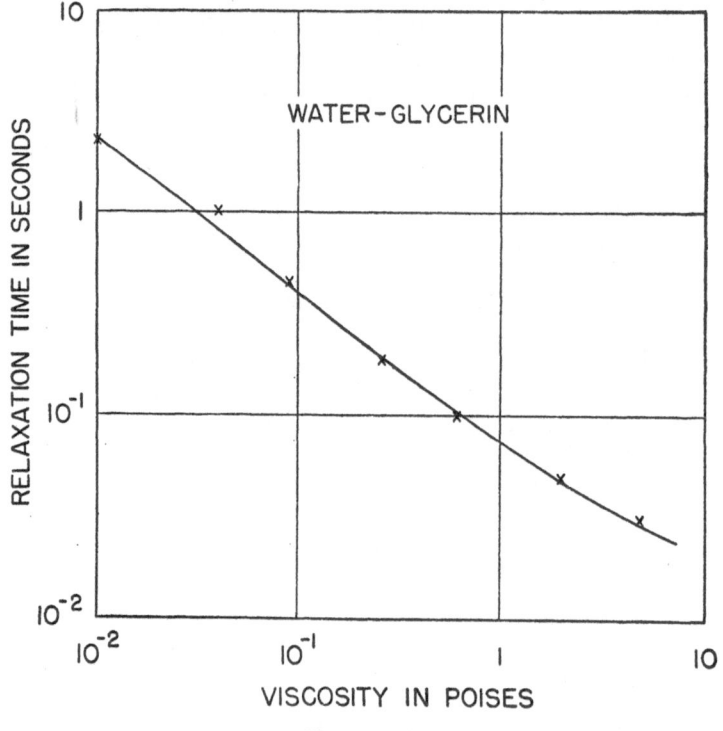

Figure 4. 1.

The relaxation time of the proton resonance at 29 Mc/sec in mixtures of water and glycerin.

diameter a changes of course, but this would cause a deviation from the inverse proportionality with η in the direction opposite to that observed. We can only say that our treatment of a molecule as a sphere with a magnetic moment in the centre becomes very crude for large molecules, each containing several protons. In the modern theory of the viscosity a concept exists, that continually transitions are made between configurations around a given molecule, which are more or less stable. The rate at which these changes in configurations take place determines our correlation time τ_c which will depend therefore in a complicated manner on the shape and size of the molecule. For the large chain-like molecules in the hydrocarbons one has furthermore the possibility of bending and twisting of a molecule, which changes the relative position of the protons in that molecule.

The reader may be reminded that similar difficulties arise in D e b y e's theory of dielectric dispersion. His time τ determined experimentally, does not always correspond to the one calculated from (4. 14). Attempts have been made to explain this deviation by taking into account the electric dipole interaction between the polar molecules and introducing different models for the electric local field. Note that glycerin which shows the largest deviation in our case, also violates D e b y e's formula (4. 14) most severely. We want to stress, however, that the D e b y e time τ and our correlation time τ_c characterize different physical processes. D e b y e's τ refers only to the orientation of the polar group in space, while for τ_c any relative reorientation between the magnetic nuclei must be considered. The following formulation then seems appropriate. The characteristic time τ of D e b y e and the correlation time τ_c in the magnetic local field spectrum are proportional in one sample. They both vary in proportion to η/T, if the temperature of the sample is changed.

The proportionality constant between τ and τ_c varies from substance to substance, depending on the detailed picture of the molecular motion in each substance, but the ratio will always be of the order of unity. For the model of a sphere in a viscous medium we have $3\,\tau_c = \tau$. Experimental values for the proportionality factor are given in section 4. 3. 1.

We can obtain a better test of the theory if we carry out measurements of the relaxation time and line width in one substance at various temperatures. We shall first describe in some detail the behaviour of T_1 and T_2, that must be expected from theory. Substitution of (4. 9) and (4. 12) into (2. 54) and (2. 53) leads to

$$1/T_1 = K_1\left[\frac{\tau_c}{1 + 4\,\pi^2\,\nu^2\,\tau_c^{\,2}} + \frac{2\,\tau_c}{1 + 16\,\pi^2\,\nu^2\,\tau_c^{\,2}}\right] \tag{4.21}$$

$$1/T_2' = \sqrt{K_0\int_{-1/\pi\,T_2'}^{1/\pi\,T_2'}\frac{\tau_c}{1 + 4\,\pi^2\,\nu^2\,\tau_c^{\,2}}\,d\nu} = \sqrt{\frac{K_0}{\pi}\,\mathrm{arc\,tg}\,\frac{2\,\tau_c}{T_2'}} \tag{4.22}$$

with $$K_1 = {}^2/_5\,\gamma^4\,\hbar^2\,I\,(I+1)\,b^{-6} \tag{4.23}$$

and $$K_0 = 3\,K_1 \tag{4.24}$$

It has been assumed that the averaging over ϑ could be carried out independently. Use has been made of the relations (4.12). Furthermore the formulae are written for a single relaxation time $\tau \sim c\,\eta/T$. Actually we have a distribution of relaxation times as we have seen for the translational effect in water. We should write instead of the constant c the function $c(\lambda)$ and integrate over the parameter λ. In most cases the distribution will be narrow, since only the nearest neighbours contribute strongly. Strictly speaking the constants K are functions of the temperature, as they vary with the density of the sample, but this effect

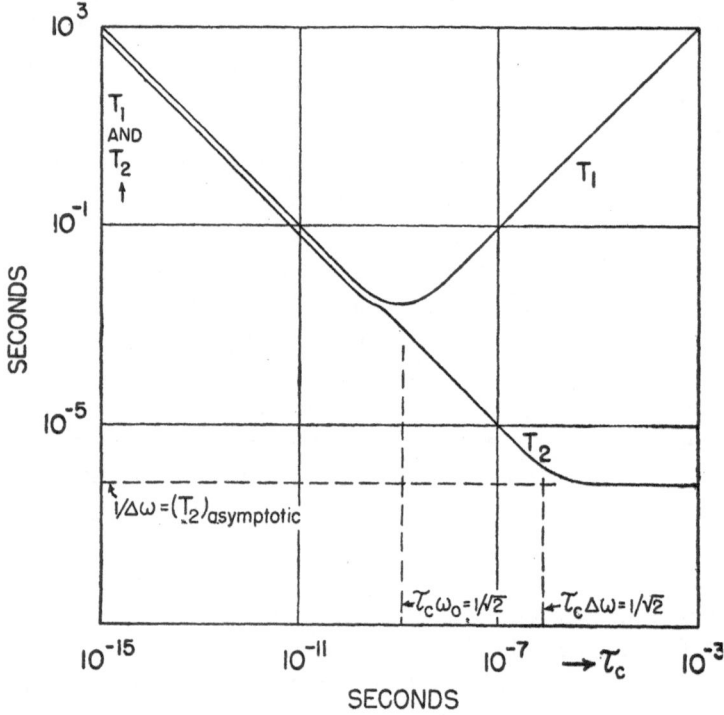

Figure 4.2.

The theoretical behaviour of the relaxation time T_1 and T_2, which is a measure for the inverse line width.

is completely negligible. The simplifying assumptions now permit to point out clearly the general behaviour of T_1 and T_2, which are plotted as a function of τ_c in fig. 4.2. Here T_2 is defined by (2.58).

For $4\pi^2\nu^2\tau_c^2 \ll 1$, T_1 is inversely proportional to τ_c and thus to η/T, and for $4\pi^2\nu^2\tau_c^2 \gg 1$ directly proportional. The plot on a double

logarithmic scale therefore shows two straight lines making angles of 45°
and 135° degrees with the x-axis.

In the transition region $(4\pi^2\tau_c^2\nu^2 \approx 1)$ T_1 has a minimum value

$$(T_1)_{min} = {}^4/_3 K_1 \tau_c \tag{4.25}$$

for $$\tau_c = {}^1/_2 \sqrt{2}\,\pi\,\nu_0$$

The quantity T_2' is a monotonic decreasing function of τ_c and reaches
an asymptotic value

$$(1/T_2')_{asymptotic} = \sqrt{{}^1/_2\,K_0} \tag{4.26}$$

for very long correlation times. This value is of course exactly the same
as the one we calculated for the static case (2.36) where the nuclei are
at rest. For $\tau_c \ll (T_2')_{asymptotic}$, T_2' is inversely proportional to τ_c. The
horizontal distance between the points, where T_1 and T_2 bend over
respectively, is given by the ratio $2\pi\nu_0\,(T_2')_{asymptotic} \approx H_0/H_{loc}$. For
$4\pi^2\nu^2\tau_c^2 \ll 1$, T_1 and T_2' are proportional and from (4.21) and (4.22)
we find for the proportionality constant

$$T_2' = {}^1/_2\,\pi\,T_1 \tag{4.27}$$

The line width is given by (2.58) with one of the relations (3.22) or
(3.23). For $4\pi^2\nu_0^2\tau_c^2 \gg 1$ we have

$$T_2 \approx T_2' \tag{4.28}$$

for $4\pi^2\nu_0^2\tau_c^2 \ll 1$ we have with (4.27)

$$T_2 = 0.85\,T_1 \tag{4.29}$$

We must not attach too much weight to this particular ratio, for about
the limits in the integral in (4.22) we only know that they must be of the
order of magnitude of the line width expressed in cycles/sec. It might be
better to take the limits as $\pm\,1/\pi T_2$ instead of $\pm\,1/\pi T_2'$. This would
not make any difference for long τ_c's, and does not affect the order
of magnitude for the region where T_1 and T_2 are proportional. We shall
see in the next paragraphs that the experimental ratio between T_1 and T_2
is close to the value predicted by (4.28) and (4.29). On this basis the
resonance line in water e.g. with $T_1 = 2.3$ sec. should be very narrow
indeed. The width should be of the order of one cycle or about 10^{-4}
oersted. The experimental width is then, of course, determined by the
inhomogeneity in H_0 as we have already pointed out several times.

4.1.4. *Experimental results in ethyl alcohol and glycerin between* + 60° C *and* —35° C.

In order to vary the temperature of the sample in the radiofrequency coil, copper tubing (3mm inside diameter) was soldered around the grounded shield of the radiofrequency coil (see fig. 3.8). To obtain low temperatures acetone, cooled by dry ice, could flow through the tubing from a container, which was placed above the magnet, under the influence of the gravitational force. This acetone was not in direct contact with the dry ice. For dissolved CO_2 would be set free, when the acetone was warmed up in passing through the narrow tubing. This would prevent a regular flow. The apparatus in the magnet gap and all other cold parts were thermally insulated with glass wool and asbestos paper. The temperature was measured by a copper-constantan thermo-element. One contact point was brought in the liquid through the small cork stop closing the thin walled glass tube which contained the sample. There was no trouble of pick-up of radio frequencies, since the coupling between the leads of the thermo-element and the coil was very small indeed, as the contact point was kept well outside the volume of the coil. The other contact of the element was put in melting ice. The thermo — E.M.F. was measured with a Leeds & Northrup type K potentiometer. The element was calibrated at + 100° C, 0° C and —78° C, which checked with the calibration data given in the Handbook of Chemistry and Physics, so that this table was used. The temperature of the sample could be varied by changing the flow of the cooling liquid. The temperature remained constant to within 0.5° C during the determination of each saturation curve. The balance of the bridge was also stable, once thermal equilibrium had been established. To cover the range of higher temperature, the container was filled with iced water or hot water.

The variation of the viscosity with temperature was taken from the Physikalisch-Chemische Tabelle. The data obtained with ethylalcohol at two frequencies are shown in fig. 4.3. The variation of the relaxation time with viscosity is inversely proportional. The line drawn through the points makes an angle of 135° with the x-axis. Although the variation in the viscosity is not large, the points clearly indicate the theoretical behaviour, to be expected for short τ_c. The real line width could not be measured. The limit set by the inhomogeneity of the field is 0.015 oersted at 4.8 Mc/sec. According to theory the line width should be much narrower than this. As was pointed out in chapter 3, any systematic errors in the relative determination of T_1 cancel out in this case. More interesting are the results for glycerin shown in fig. 4.4. The freezing point of this substance is 18° C, but it usually gets supercooled and very high viscosities are obtained at low temperature, where the substance becomes almost glasslike. The experimental points show that we have reached the region where $2 \pi \nu_0 \tau_c > 1$. The drawn lines are theoretical curves. The observed minima are somewhat flatter and on the low temperature side the points do not quite fit a 45° line. This can, at least in part, be explained by a distribution of correlation times τ_c, rather than the single value to which

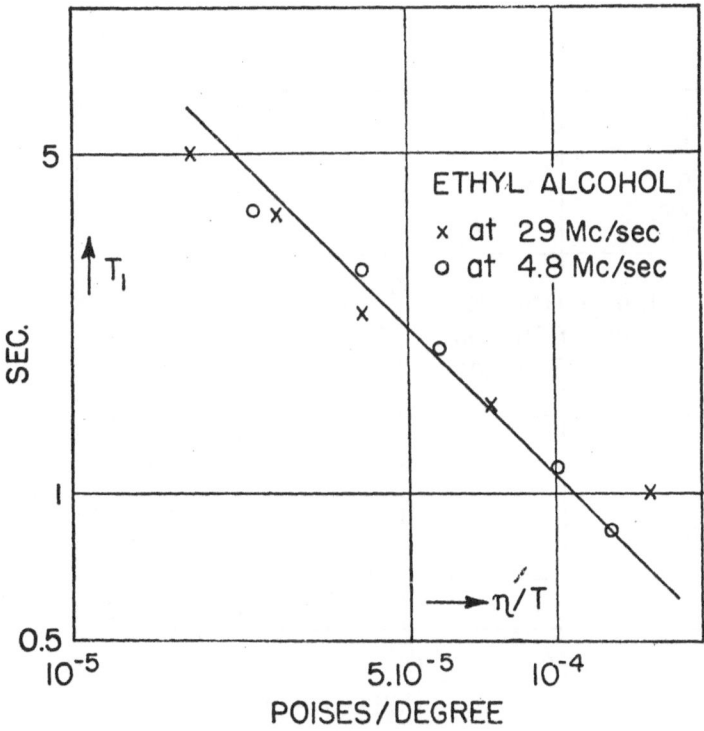

Figure 4. 3.

The relaxation time of the proton resonance in ethyl alcohol between 60° C and — 35° C. The straight line makes an angle of 45° degrees with the negative x-axis.

the theoretical curves pertain. It would be interesting to extend the measurements to lower temperatures to get more information about this distribution. The shift of the minimum with frequency is somewhat less than predicted by (4. 25). We find a factor 4 instead of 6. On the low temperature side the relaxation time should be proportional to ν_0^2. Instead of a factor 36 we find a factor 14. Again this deviation can, at least partly, be understood by remarking that (4.25) holds only in case of a single correlation time, or if one wishes, of a single correlation function. The data on the line width are plotted in the same diagram with the aid of formula (3.22) for a Gaussian curve.

At room temperature the line is narrower than the inhomogeneity of the external field. Extrapolation of the dotted line towards higher temperatures gives the ratio $T_1/T_2' = 1$. In the region where T_1 is proportional to the viscosity and T_2 inversely proportional, the saturation of the line

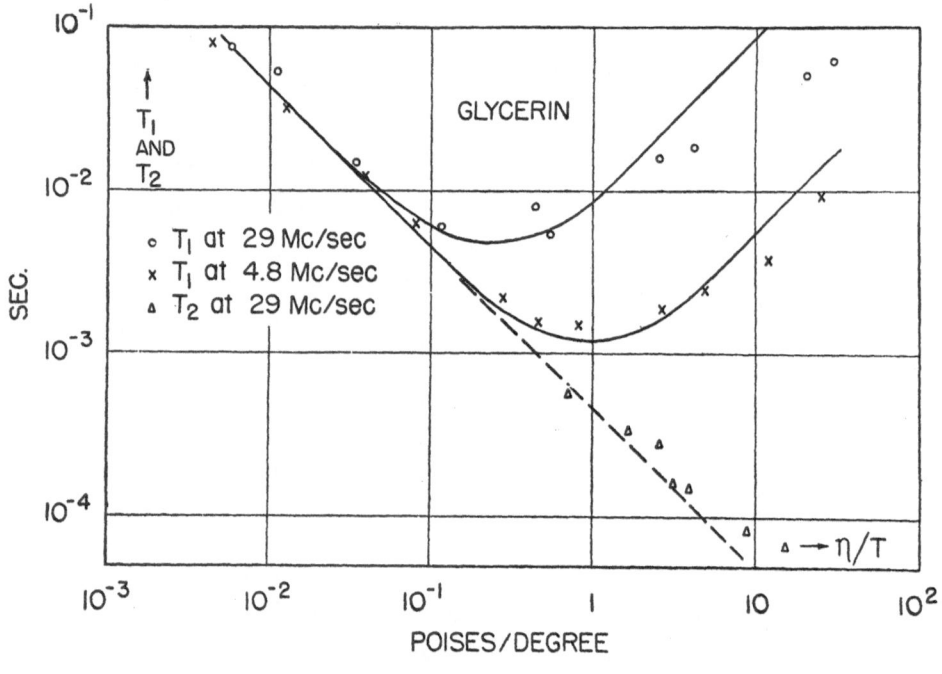

Figure 4.4.

The relaxation time and the line width of the proton resonance in glycerin between 60° C and −35° C. The lines, drawn through the experimental points, have the theoretical form of fig. 4.2.

always occurs at the same output power of the generator, that is at the same density of the applied radio frequency field, as the product T_1T_2 is constant. From the viscosity, measured at 20° C, it followed that the glycerin used in the experiment was not pure and probably contaminated with 2 % water. Experiments carried out with mineral oil gave similar results both for the relaxation time and line width.

4.1.5. *The influence of paramagnetic ions.*

So far we have considered the dependence of the relaxation time on τ_c. It is also possible, however, to bring about changes in the quantities K_1 and K_0 in (4.21) and (4.22) by mixing the substance with paramagnetic ions. From (4.23) and (4.24) we see that the large γ-values of the electronic moments will enhance the values of K_1 and K_2. The larger interaction of the nuclear moment with the electronic moment will shorten the relaxation time and enhance the line width, τ_c remaining constant. Let us consider an aqueous solution of ferric nitrate. We can calculate the influence of the Fe^{+++} ions in the same way as we did, when we estimated the contribution of the protons in other molecules to the relaxation time in pure water. An adapted formula (4.20) would read

$$1/T_1 = 12\,\pi^2\,\gamma_p{}^2\,\gamma_{\text{ion}}^2\,\hbar^2\,S_{\text{ion}}\,(S_{\text{ion}}+1)\,N_{\text{ion}}\,\eta/5\,k\,T \qquad (4.30)$$

This applies for ions of the iron-group, which are of the "spin-only" type. For others we should replace $\gamma_{\text{ion}}^2\,\hbar^2\,S_{\text{ion}}\,(S_{\text{ion}}+1)$ by μ_{eff}^2.

Of course we should add to (4.30) the contribution of the protons in the solution, which in pure water are solely responsible for the relaxation time. But as γ_{ion}^2 is about 10^6 times larger than $\gamma_p{}^2$, the influence of the paramagnetic ions is predominating even in a concentration of 10^{-3} N. or 10^{18} ions/cc. According to (4.30) the relaxation time should be inversely proportional to the concentration and to the square of the magnetic moment of the paramagnetic ions. In fig. 4.5 the results for three ions are given. It appears that the curves, also to the absolute magnitude, can be well represented by (4.30). Only for very low frequencies there seems to be a deviation towards longer relaxation times. This is all the more remarkable since the straight lines finally must bend over to the left to the asymptotic value of 2.3 sec. in pure water. We do not know if the effect is real. It certainly seems too big for a systematic error. We would like to point out that (4.30) certainly needs some correction. For while the motion of a watermolecule relative to the ion is still given by (4.17) and (4.18), where a is the radius of the watermolecule, the distance of closest approach is determined by the radius of the ion and its hydratation. We must insert a correction factor a/b. It is very hard to estimate correctly the motion of a watermolecule in the dipole atmosphere around an ion. But if there is an effect from the hydratation, it should become more pronounced at small concentrations.

Furthermore we should take into account that the correlation time in the local field spectrum is not solely determined by the molecular motion in the liquid, but also by changes in quantisation of the electronic spins,

which possibility was already indicated in (2.42). The characteristic
time for this latter process is not known experimentally, as the paramagnetic electronic relaxation times $\varrho/2\pi$ in solutions are short, of the
order of 10^{-10} sec.[1]). This implies that in the derivation of (4.30) we
should have used for τ_c the constant ϱ instead of (4.18) for values of

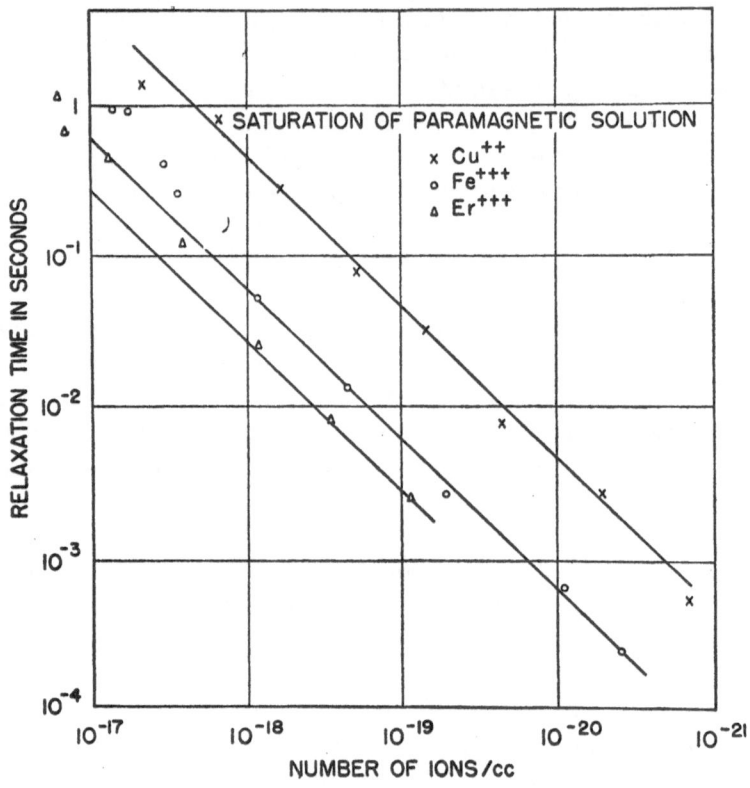

Figure 4.5.

The relaxation time of the proton resonance in aqueous solutions of
paramagnetic salts. The lines, drawn through the experimental points,
make angles of 45° with the negative X-axis.

r, where τ_c would become larger than ϱ. This reduces only the influence
of the ions which are rather far away, so that this correction is not

[1]) One might be tempted to calculate ρ in the same way as we did for the nuclear
relaxation time. However, more important than the magnetic interaction between the
spins will be the electric interaction in the polar liquid via the spin-orbit coupling. The
only experimental information, known to the author, comes from Z a v o i s k y (Z 1).

important. The inverse proportionality with μ_{eff}^2 is rather well realised for some ions, and completely violated for others (Ni^{++}, and especially Co^{++} and $Fe(CN)_6^{---}$), as is shown in Table III.

TABLE III.

Ion	μ_{eff} in Bohr magnetons from relaxation experiments	μ_{eff} in Bohr magnetons from susceptibility measurements
Er^{+++}	9.5	9.4
Fe^{+++}	6.3	5.9
Cr^{+++}	4.7	3.8
Cu^{++}	2.3	1.9
Ni^{++}	2.1	3.2
Co^{++}	1.3	4.5—5.3
$Fe(CN)_6^{---}$	0.12	2.4

The second column is computed with (4. 30) from measurements of the nuclear relaxation time in solutions of known concentration. The values in the last column were taken from G o r t e r (G 3).

They were obtained from the measurement of the static susceptibility of solutions (comp. V 1). The value for $Fe(CN_6)^{---}$ was taken from measurements on solid $K_3Fe(CN)_6$ (J 1). The large deviations for the last three ions can be understood, because nondiagonal elements [1] contribute greatly to the magnetic moment of these ions. With these elements components of the local field spectrum are connected, which have a higher frequency than $v_0 + 1/\tau_c$, where $1/\tau_c$ is the limit where the local spectrum caused by the Brownian motion drops off rapidly. Thus these non-diagonal elements do not contribute to the nuclear relaxation mechanism, and the μ_{eff} for this process is correspondingly smaller. The extremely small influence of $Fe(CN)_6^{---}$ is probably partly caused by the six CN groups around the iron atom, so that the b is very large. For variations of b for the various ions have not been taken into account in Table III.

Finally we may ask what the influence can be of oxygen gas dissolved in water. The magnetic moment of O_2 is 2.8. The maximum concentration of dissolved O_2 in water at room temperature under 18 % of the atmospheric pressure is 1.5×10^{17} molecules/cc. The relaxation time, due to O_2 alone, could not be smaller than 2.5 sec. The relaxation time in water is therefore determined by the neighbouring protons and the dissolved

---

[1] For Co^{++} and $Fe (CN)_6^{---}$ even important deviations from C u r i e's law have been found.

oxygen. In the determination of the absolute value of the relaxation time (see chapter 3) distilled water was used. As the distillation was not done in vacuo, we have no guarantee that for pure water the relaxation time is not somewhat longer.

We now consider the line width in the solutions. As the correlation time τ_c in paramagnetic solutions is essentially the same as in water and thus $4\pi_0\nu_0^2\tau_c^2 \ll 1$, we expect that T_2 is proportional to T_1. This is confirmed by the experimental result in fig. 4. 6. The line width, measured

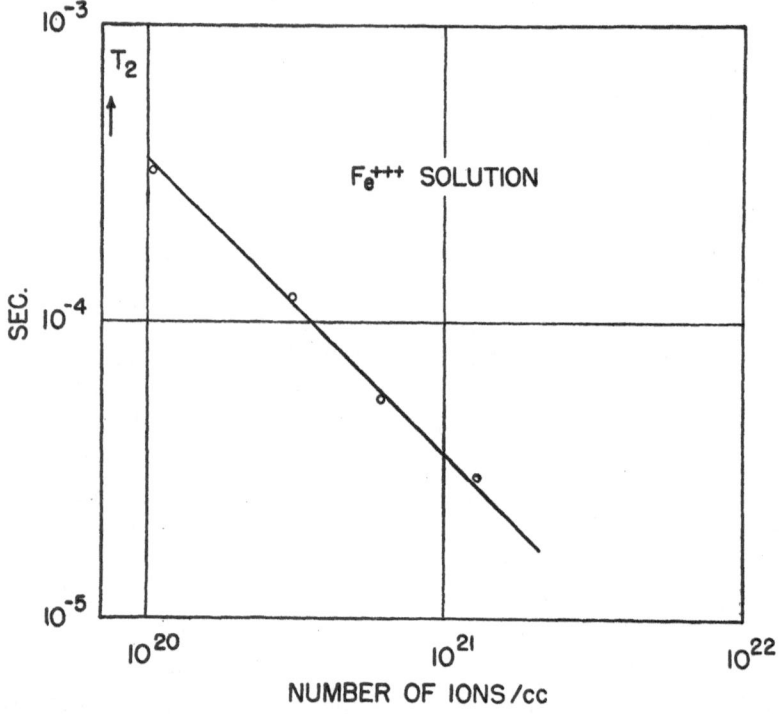

Figure 4. 6.

The line width of the proton resonance in aqueous solutions of $Fe(NO_3)_3$. The quantity T_2 is inversely proportional to the line width, which appears to be proportional to the concentration.

between the points of maximum slope in an assumed Gaussian, is $2/\gamma\,T_2$, and is proportional to the concentration. For small concentrations the width is again too narrow to be measured. Comparison of fig. 4. 5 and 4. 6 yields $T_2 = 1/2\;T_1$ or $T_2' = 2/3\,T_1$. The same ratio was found for Cu^{++} solutions and is in good agreement with the value found in glycerin.

It may be well to point out here that the proton resonance in paramagnetic solutions appears to be shifted, because the field inside the sample is different from the field elsewhere in the gap. The microscopic field inside the sample at the position of the protons always determines the position of the proton resonance. We are interested in the field produced by all paramagnetic ions at the position of a proton and not of all but one at the position of another ion. It is not permissible to put the macroscopic $\vec{H}$ inside the sample into the resonance condition (1.7). One has to take the average microscopic field at the position of the protons. At the same time we might mention another factor which changes slightly the magnetic field experienced by a nucleus, namely the diamagnetism of the surrounding electrons. This effect has been calculated by R a b i and coworkers and is very small for light elements (K 12).

4.1.6.   *The resonance of F^{19} and Li^7 in liquids.*

To compare the resonances of F^{19} and H^1 in a liquid compound, a "Freon", $CHFCl_2$, monofluoro-dichloro-methane, was condensed in a glass tube and sealed off. Both the H^1 and F^{19} resonance were narrower than the inhomogeneity in the field. The total intensity of the two lines was the same (within 15 %) so that it was confirmed that F^{19} has the same spin as the proton. The relaxation times were 3.0 sec. for H^1 and 2.6 sec. for F^{19}. The γ_F is 6.5 % smaller than γ_P, but the F^{19} nucleus experiences a somewhat larger local field as its nearest neighbour is the proton in the same molecule, while the proton has in turn the F^{19} nucleus. We should expect on this basis the relaxation times to be the same, as is confirmed within the experimental error.

Experiments were also carried out in solutions of KF. Since the signal to noise ratio drops proportional to the number of nuclei per cc, only very concentrated solutions could be investigated to obtain a sufficiently intense F^{19} resonance. Again the resonance lines are narrow. The result for the relaxation times is shown in fig. 4.7. The decrease in the proton relaxation time can be explained by the increase in viscosity of the concentrated solution. The much more pronounced decrease for fluorine may be an indication that the motion of these ions is more quenched, when one comes very close to the transition point, where the solution changes into the solid hydrate $KF.2H_2O$. A more careful study of the nuclear relaxation might give information about the character of this, and other, transitions. Anticipating the results for solids we can say that in the crystalline $KF.2H_2O$ the lines are wide and that we are in the region where $4\pi^2 \tau_c^2 \nu_0^2 \gg 1$.

An interesting substance is also BeF_2, which can be mixed with water in any proportion. For high concentrations the substance becomes very viscous, and finally goes over into the glasslike, amorphous BeF_2, when no water is present. Preliminary experiments showed that the behaviour of both the proton and the fluorine resonance in $Be F_2 + H_2O$ is similar to that of the proton resonance in glycerin. With increasing viscosity of the mixture the relaxation time first drops to about 10^{-3} sec., then rises

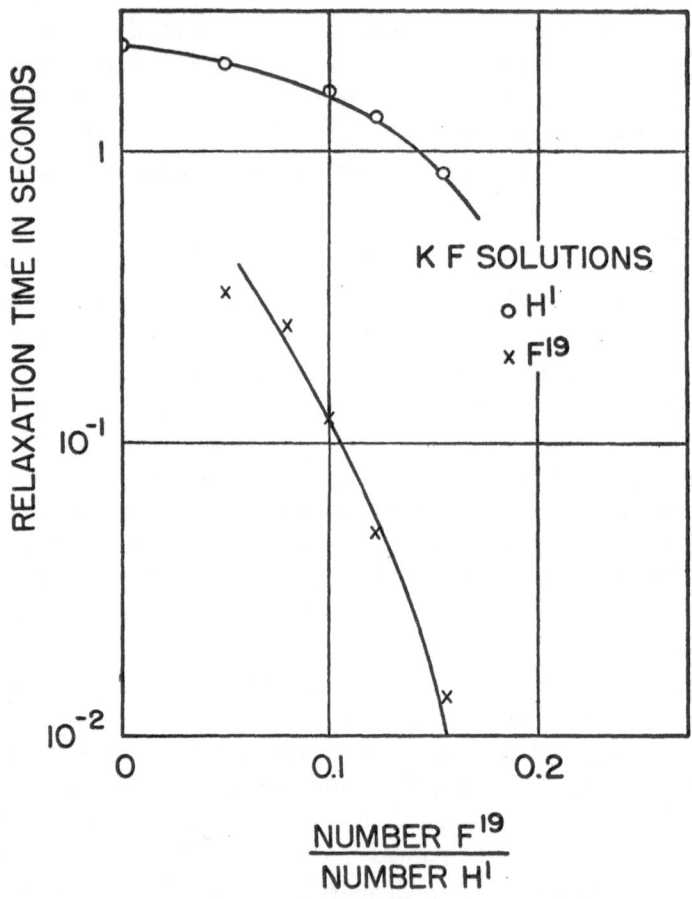

Figure 4. 7.

The relaxation time of the proton and fluorine resonance in aqueous solutions of K F of various concentrations.

again to 0.2 sec in pure BeF_2. The line width measured between the points of maximum slope increases from very small values to about 10 oersted in pure BeF_2.

Experiments on the Li^7 resonance were carried out at 14.5 Mc/sec

TABLE IV.

Substance	Number of Li atoms	Relaxation time in seconds	
	Number of H atoms	for Li^7	for H^1
$LiCl + H_2O$	5.8	1.75	0.4
$LiNO_3 + H_2O$	14	2.7	1.1
$LiNO_3 + H_2O + Fe(NO_3)_3$	8	0.11	0.002³
$LiCl + H_2O + CrCl_3$	6.5	0.24	0.009⁵
$LiCl + H_2O + CuSO_4$	6.6	0.18	0.01⁸

in solutions of LiCl and $LiNO_3$. Table IV gives some results. For the solutions without paramagnetic ions the decrease in relaxation time of the proton resonance compared to pure water can be explained by an increase in viscosity of the concentrated solutions. The relaxation time for Li^7 in this case is somewhat longer. The ratio of the local field spectra is given by

$$\frac{\text{Spectral intensity at } Li^7 \text{ nucleus}}{\text{Spectral intensity at proton}} = \frac{(T_1 \gamma^2)_p}{(T_1 \gamma^2)_{Li^7}}$$

Since $\gamma_p^2 / \gamma_{Li}^2 = 6.6$, the local field has a somewhat higher intensity at the Li-nucleus. The cause could be the slower motion of the largely hydrated Li-ion. A more likely explanation however, is, as we shall see later, that the intensity of the magnetic local field is the same, or even smaller but that there is a contribution to the relaxation process from the quadrupole moment of Li^7, which has a spin $I = 3/2$.

The influence of paramagnetic ions is much smaller on Li^7 than on the protons. In the first place the local field spectrum at the Li^7 nucleus will be smaller because the repulsion of two positive ions will make it less likely for them to come close together, and then they have to compete with the quadrupole transitions (cf. chapter 5). At the conclusion of this paragraph we direct the attention of the reader to the results found by other investigators (B 2, B 7, R 5), which seem to be in agreement with the general ideas, here proposed. Especially we might mention the experiment in liquid hydrogen by R o l l i n (R 6).

4. 2. *The relaxation time and line width in gases.*

4. 2. 1. *Hydrogen.*

The only experiment of nuclear magnetic resonance (P 6) in gases which has been reported was performed with hydrogen gas at room temperature between 10 and 30 atmospheres of pressure. The accuracy was poor, as the density of the nuclei is low. It was found that the line is narrow (< 0.15 oersted) and that the relaxation time at 10 atmospheres $T_1 \approx 0.015$ sec. with an indication that T_1 increases with increasing pressure. We shall now investigate what the theory predicts for this case.

The local field at the position of a proton in an H_2-molecule in a volume of hydrogen gas consists in the first place of the contribution connected with the rotational moment $\vec{J}$ of the molecule and the magnetic moment of the other proton. According to P a u l i's exclusion principle the spins of the two protons can only be parallel, if the electronic wave function is antisymmetric (J odd, orthohydrogen), and only anti-parallel, if the electronic wave function is symmetric (J even, parahydrogen). The transitions from the ortho- to the para-state in hydrogen gas are extremely rare. Furthermore, if the system is in thermal equilibrium at room temperature, 13 % of the H_2 molecules have $J = 0$, 66 % have $J = 1$, 12 % have $J = 2$ and 9 % have $J = 3$. We ignore for the sake of the simplicity transitions from $J = 1$ to $J = 3$. We assume that the rotational angular momentum of orthohydrogen is a constant of the motion. The total nuclear spin $I = I_1 + I_2$, $I = 0$ for parahydrogen, $I = 1$ for orthohydrogen. Only orthohydrogen will show nuclear resonance. At room temperature equilibrium the ratio of molecules in ortho- and para-states is as $3 : 1$. So the total intensity of the nuclear magnetic absorption line is proportional to $\frac{3}{8} NI(I + 1)$. This is equal to $(\frac{1}{2} \cdot \frac{3}{2})N$. Thus the total intensity of the line of orthohydrogen is the same as if all N protons were uncoupled in hydrogen atoms.

If the molecule is placed in a strong magnetic field, in zero approximation not only I and J, but also m_I and m_J are constants of the motion. We first consider the interaction of the nuclear spin with the rotational moment. The perturbation term in the Hamiltonian is given by

$$H_{op} = \gamma \hbar H' \vec{I} \cdot \vec{J}$$

$$= \tfrac{1}{2} \gamma \hbar H' \{(I_x + iI_y)(J_x - iJ_y) + (I_x - iI_y)(J_x + iJ_y)\} + \gamma \hbar H' I_z J_z$$

$$(4.31)$$

From R a b i's experiments (K 3) follows the value of H'; the magnetic field at the position of the protons produced by the rotation of the molecule is 27 oersted. With (4. 31) we can once more repeat the reasoning explained in sections 2. 4 and 2. 5 in order to calculate the relaxation time. If the quantisation of J were fixed, that is if m_J did not change during collisions, we would have no transitions in m_I. For the first two terms on the left hand side in (4. 31), which have non-diagonal elements in m_I, involve also a change in m_J. But the collisions in the gas will change m_J and we can assume that after each collison m_J has equal chance for any of its $2J + 1$ values. As the distribution of the collisions in time of a given molecule, measured from the time of the preceding collision, is given by $\frac{1}{\tau_c}\exp{-t/\tau_c}$, where τ_c is the mean collision time in the gas, we have a Fourier spectrum for m_J and thus for $J_x - i J_y$. The intensity of the spectrum of the latter is with (4. 9)

$$J(\nu) = \frac{4 J(J+1)}{3} \frac{\tau_c}{1 + 4 \pi^2 \nu^2 \tau_c^2} \tag{4. 32}$$

From this and (4. 31) we obtain a relaxation time

$$1/T_1 = \frac{2 \tau_c}{1 + 4 \pi^2 \nu_0^2 \tau_c^2} H'^2 \gamma_p^2 \frac{J(J+1)}{3} \tag{4.33}$$

with
$$\tau_c = 1,4/ v \sigma N \tag{4. 34}$$

The number of molecules per cc, proportional to the pressure, is denoted by N, σ is the collision cross section, v is the average velocity of the molecules.

To (4. 33) we have to add the contribution of the spin-spin interaction, which is represented by the perturbation term

$$H_{op}'' = \frac{-\gamma_p^2 \hbar^2}{r^3} \left[3 (\vec{I_1} \cdot \vec{n})(\vec{I_2} \cdot \vec{n}) - \vec{I_1} \cdot \vec{I_2} \right] \tag{4. 35}$$

where $\vec{n}$ is the unit vector pointing from one proton to the other, and r is the distance between them. The expression (4. 35) can be transformed to one which only contains constants and the operators $\vec{J}$ and $\vec{I} = \vec{I_1} + \vec{I_2}$,

$$H_{op}'' = \gamma_p^2 \hbar^2 \frac{1}{r^3} \frac{I(I+1) + 4 I_1(I_1+1)}{(2I-1)(2I+3)(2J-1)(2J+3)} \left[3 (\vec{I}.\vec{J})^2 + {}^3/_2 \vec{I}.\vec{J} - \vec{I^2}.\vec{J^2} \right] \tag{4. 36}$$

In order to find the contribution of this interaction to the relaxation process, we have to write the operator between square brackets in the

m_I, m_J representation. R a b i and collaborators (K 4) found that this operator is equal to

$$\tfrac{1}{2}\,[3\,J_z{}^2 - J\,(J+1)]\,[3\,I_z{}^2 - I(I+1)] + \qquad \triangle\,m_I = \triangle\,m_J = 0$$

$$+\,\tfrac{3}{4}\,[J_z\,(J_x + i\,J_y) + (J_x + i\,J_y)\,J_z]\,[I_z\,(I_x - i\,I_y) + (I_x - i\,I_y)\,I_z]$$
$$- \triangle\,m_I = +\,\triangle\,m_J = 1$$

$$+\,\tfrac{3}{4}\,[J_z\,(J_x - i\,J_y) + (J_x - i\,J_y)\,J_z]\,[I_z\,(I_x + i\,I_y) + (I_x + i\,I_y)\,I_z]$$
$$\triangle\,m_I = -\,\triangle\,m_J = 1$$

$$+\,\tfrac{3}{4}\,(I_x + i\,I_y)^2\,(J_x - i\,J_y)^2 \qquad \triangle\,m_I = -\,\triangle\,m_J = 2$$

$$+\,\tfrac{3}{4}\,(I_x - i\,I_y)^2\,(J_x + i\,J_y)^2 \qquad -\,\triangle\,m_I = \triangle\,m_J = 2$$
$$(4.\,37)$$

The matrix elements can be written down immediately with the rules of matrix multiplication and the expressions (1,1), (1,2) and (1,3). The matrix elements of (4. 37) with $\Delta m_I = 1$ and 2, combined with the components at ν_0 and $2\nu_0$ of the frequency spectrum of the corresponding terms in m_J give an expression for $1/T_1$, which must be added to (4. 33).

We write down the final result, first derived by S c h w i n g e r for the case realized in practice that τ_c is short compared to the Larmor period $1/\nu_0$.

$$(1/T_1)_{H_2 - \text{gas}} = 2\,\tau_c\,\gamma_p{}^2\left[\tfrac{1}{3}\,H'^2\,J\,(J+1) + 3\,H''^2\,\frac{J\,(J+1)}{(2J-1)\,(2J+3)}\right]$$
$$(4.\,38)$$

where $H'' = \dfrac{\overline{1}}{r^3}\,\tfrac{1}{2}\gamma\,\hbar$ is the effective field from one proton at the position of the other. From R a b i's experiments (K 3) follows $H'' = 34$ oersted. In (4. 38) we have already asumed $4\,\pi^2\,\nu_0{}^2\,\tau_c{}^2 \ll 1$. This is always fulfilled under practical conditions. The opposite case $4\,\pi^2\,\nu_0{}^2\,\tau_c{}^2 \gg 1$ would occur at pressures of 1 mm Hg or less, where the signal is much too small to be detected. From (4. 38) and (4. 34) it follows that the relaxation time T_1 is proportional to the pressure. Substituting numerical values $\gamma_p = 2.7 \times 10^4$, $J = 1$, $\tau_c = 10^{-11}$ sec. (Handbook of Chemistry and Physics) for a pressure of 10 atmospheres, we find $T_1 = 0.03$ sec, which is in agreement with the experimental value.

The line width can be calculated on similar lines as we did in chapter 2 from (4. 31) and (4. 32). As in liquids we find again that T_2 is of the same order as T_1, so that the resonance line should be very narrow. As T_2 is proportional to T_1, the line width should be inversely proportional to the pressure. We can speak of "pressure-narrowing" of the nuclear resonance line in H_2-gas.

The conclusion is: The magnetic interactions in the H_2-molecule give rise to a fine structure of the radiofrequency spectrum in R a b i's molecular beam method (K 3). Combined with the collisions in the gas sample for pressures > 10 mm Hg, as used in P u r c e l l 's method, they give rise to a relaxation mechanism and the local fields average out to a single very narrow line.

We have not considered the influence of the other molecules during a collision on the relaxation time. In the next paragraph we shall see, that this effect can usually be neglected in H_2-gas.

4. 2. 2. *Helium.*

An entirely different state of affairs occurs in the interesting case of He^3 gas. The atoms are in an S-state. The only perturbation is brought about during the collisions by the nuclear magnetic moment of the colliding atom. Unlike in hydrogen, here the influence of the other molecules is the only effect. Suppose that the He^3 nucleus has the set of eigenfunctions ψ_n in the constant field H_o. We ask for the chance that the perturbation by a collision brings the system from the initial state i with energy E_i to the final state f with energy E_f. The perturbation method, which may be applied, if the chance in one collision is small compared to unity, gives for the probability to find the system in state f after the collision

$$w_f = \frac{\sin^2 \dfrac{E_f - E_i}{\hbar} t}{(E_f - E_i)^2} \left| (f \mid H_{op} \mid i) \right|^2 \tag{4.39}$$

We cannot say precisely, what is going on during the collision. But the order of magnitude of the matrix element of the perturbation operator between the initial and final state will be the same as that of the interaction energy $\approx \gamma_1 \gamma_2 \hbar^2 d^{-3}$. The colliding particles have magnetogyric ratio's γ_1 and γ_2 and d is the distance of closest approach between the moments during the collision. The time t, during which a strong interaction takes place, is probably $\approx 10^{-15}$ sec, at any rate $t << \hbar/E_f — E_i \approx$ 10^{-8} sec. We can therefore write instead of (4. 39)

$$w_f = \gamma_1{}^2 \gamma_2{}^2 \hbar^2 t^2 d^{-6} \qquad (4.40)$$

If v is the relative velocity of the colliding particles, we have $t \approx d/v$. We then multiply by the number of collisions per second $1/\tau_c$ and find for the relaxation time

$$1/T_1 = 2 \gamma_1{}^2 \gamma_2{}^2 \hbar^2 d^{-4} v^{-2} \tau_c{}^{-1} \qquad (4.41)$$

Substituting numerical values for He³ at room temperature and atmospheric pressure, $v = 1.4 \times 10^5$ cm/sec., $\tau_c = 2 \times 10^{-10}$ sec, $d = 2 \times 10^{-8}$ cm, $\gamma_1 = \gamma_2 = 2.4 \times 10^4$, we find $T_1 = 10^6$ sec. In order to avoid saturation during the resonance measurements it is therefore necessary to admit oxygen gas. The magnetic moment of an O_2 molecule is about 10^3 times as large as of a He³ atom.

Taking $\gamma_1 = 2.4 \times 10^4$, $\gamma_2 = 2.8 \times 10^7$, $d = 2.5 \times 10^{-8}$ cm, $\tau_c = 10^{-10}$ sec we find for the relaxation time of He³ resonance, if the partial pressure of the oxygen is one atmosphere, $T_1 \approx 1$ sec. From (4.41) and (4. 35) it follows that in this case the relaxation time is inversely proportional to the pressure. Strictly speaking we ought to add a term which is similar to (4. 41) to (4. 38) in the case of H_2. From the order of magnitudes, resulting from (4. 38) and (4. 41), we see that such a term in pure H_2 gas is completely negligible for pressures below 10^3 atmospheres. For O_2 pressures of 10^2 atmospheres, however, it is an important contribution. In general we can say that most gases, consisting of molecules, will behave like hydrogen and show the "anomalous" pressure-narrowing. The noble gases, consisting of atoms in an S-state, will behave like He³ and have pressure broadening.

We shall now derive the relation between T_1 and T_2 for the case of He³. At the same time we obtain an independent derivation of the saturation formula (2. 64). The He-nuclei can be considered as completely free most of the time, but during each collision there is a small chance for the nucleus to change its orientation. The probability $w = \frac{1}{2}T_1$ for such a transition is given by (4. 40). If a radio frequency field H_1 is switched on at $t = 0$, the free nuclei will oscillate between the upper and lower state according to Rabi's formula (2. 11), until the situation is interrupted by a thermal transition. We start out with the system of nuclei in thermal equilibrium. The situation can be described by the number of surplus nuclei, originally $+ n_o$ in the lower state, oscillating between $+ n_o$ and $-n_o$, while the collisions tend to restore the equilibrium value $+ n_o$. The probability that this is achieved in the time interval between t and $t + dt$ is given by $1/T_1 \exp(-t/T_1)\, dt$, as T_1 is the average time and the distribution of gas kinetic collisions in time is given by an exponential. The average energy dissipated from the spin system and absorbed during the collisions is

$$ n_0\, h\, \nu_0 \int_0^\infty w_{1/2,\, -1/2}\, \frac{1}{T_1}\, e^{-t/T_1}\, dt \qquad (4.42) $$

where $w_{1/2,-1/2}$ is the probability that the surplus nuclei are in the upper state at time t (2.11).

After the equilibrium has been restored, the process repeats itself. In our description we have artificially broken up the natural process into self repeating steps. In reality the individual nuclei each have a chance to make transitions both up and down, with a preference for the latter. The energy absorbed per second, which must be supplied by the radio frequency field, is obtained if we multiply (4.42) by $1/T_1$, the number of times that the process is repeated per second. The integration over t can be evaluated by partial integrations. The absorbed power P is given by

$$ P = n_0\, h\, \nu_0^3\, \sin^2\vartheta\, \frac{2\,\pi^2\, T_1}{1 + 4\,\pi^2\, T_1^2\,(\nu^2 + \nu_0^2 - 2\nu\,\nu_0\cos\vartheta)} \qquad (4.43) $$

Since always $H_1/H_o \ll 1$, we can put $\sin\vartheta \approx H_1/H_o$ and $\cos\vartheta \approx 1 - H_1^2/H_o^2$, $\gamma = 2\pi\nu_0/H_0$. Near resonance $\nu \approx \nu_0$ we then have

$$ P = {}^1/_4\, n_0\, h\, \nu_0\, \gamma^2\, H_1^2\, \frac{2\, T_1}{1 + T_1^2\,(\omega - \omega_0)^2 + \gamma^2\, H_1^2\, T_1^2} \qquad (4.44) $$

which on comparison with (2.71) and (2.66) appears to be the B l o c h formula with $T_1 = T_2$.

It is interesting to apply the noise formula (3.21) to the case of He^3 and see what the minimum detectable amount is. It is not justified, however, to put in that formula $T_1 = T_2$, since the line width will always be determined by the inhomogeneity in the field. Using 10 atmospheres of O_2 we have $T_1 = 10^{-1}$ sec and we can take $T_2/T_1 \sim 10^{-2}$.

Substituting for q, λ and F each $\frac{1}{2}$ of their ideal values of unity and taking $Q = 10^2$, $\gamma = 2.4 \times 10^4$, $H_o = 10^4$, we find that 1 cc of He^3 gas at room temperature and atmospheric pressure would give a signal to noise ratio about 5, if the indication time of the meter is one second. In practice it would be very hard to find such a signal of such an extremely narrow line. One would have a better chance by searching for the moment in liquid He³ at 1° K.

Added in the proof:

Very recently A n d e r s o n (A 5) succeeded in measuring γ_{He^3} in a mixture of He³ and O_2, each at a partial pressure of 10 atmospheres.

4. 3.   *The relaxation time and line width in solids.*

4. 3. 1.   *Solids, to which the theory for liquids is applicable.*

In some solids there seems to be sufficient freedom of motion (S 5)

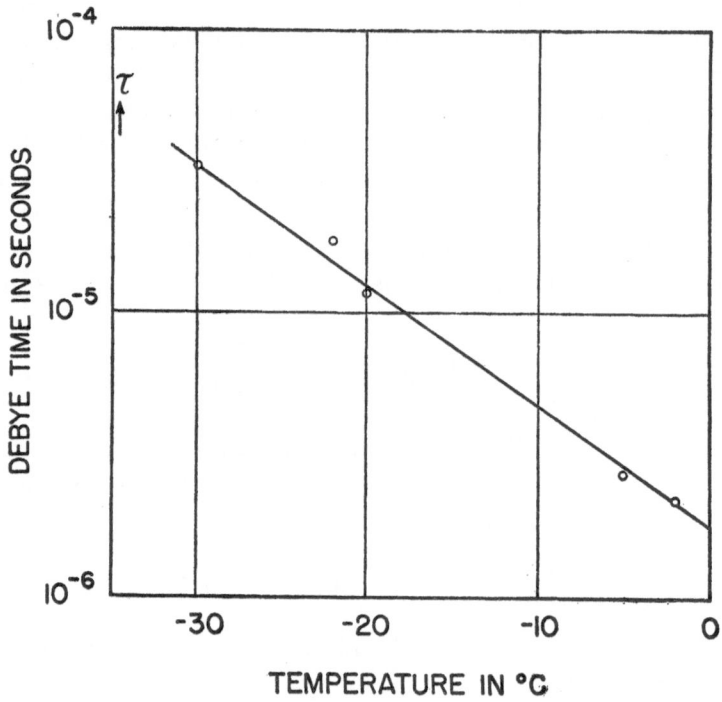

Figure 4. 8.

Values of the dielectric relaxation time τ defined by Debye, in ice at various temperatures. The points, indicated in the graph, are obtained from measurements of the anomalous dielectric dispersion in ice by Wintsch.

for the particles, that we can apply the same theory as in liquids. This state of affairs was already evident from the dielectric dispersion of the D e b y e type occurring in solids (D 2). The typical example is ice, of which we show the D e b y e time τ as a function of temperature in fig. 4.8.

The data are calculated from measurements by W i n t s c h (W 5).
Of course, the molecules are not as free as in water; τ is about 10^6
times larger than in water. We expect then that the correlation
time τ_c in the local field spectrum has increased by about the same
factor, so that the relaxation time in ice will behave in the same way as
in glycerin at low temperatures where $4\,\pi^2 \nu_0^2 \tau_c^2 \gg 1$. In fig. 4. 9 T_1
in ice between —2° C and —40° C is shown as a function of the Debye

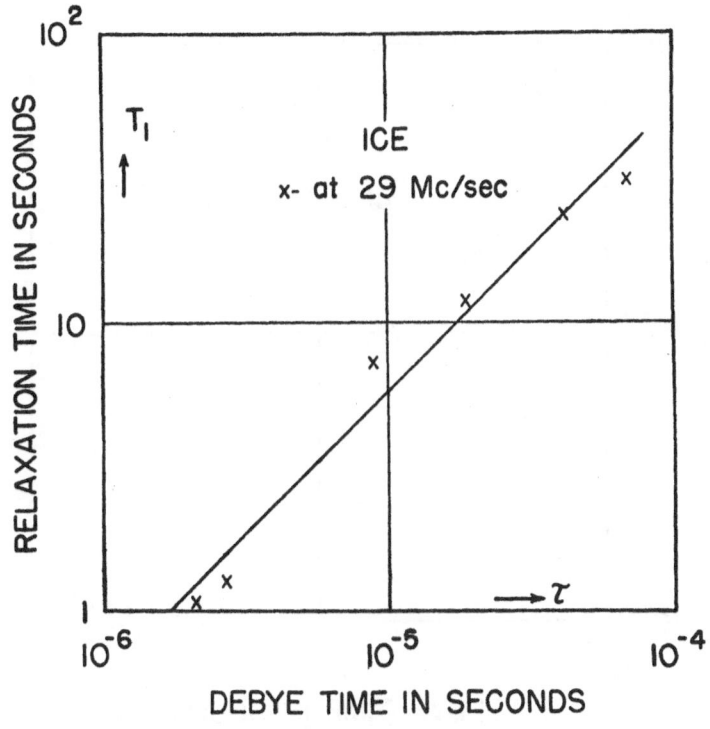

Figure 4. 9.

The relaxation time of the proton resonance in ice between — 2° C
and — 40° C, plotted against the Debye time τ. The line drawn
through the experimental points, makes an angle of 45° with the
positive X-axis.

time, to which τ_c is proportional. The graph apparently confirms the
ideas set forth in the beginning of this chapter. The straight line drawn
through the points makes an angle of 45° with the x-axis. Unfortunately
we were not able to investigate the resonance in ice at 4.8 Mc/sec,
because the signal to noise ratio became too low in that case. We would

expect, of course, the relaxation time to be shorter, but having the same dependence on τ.

Measurements of the line width yield values of T_2, which are shown in fig. 4.10. The drawn line is the theoretical curve computed from (4.22). So here τ_c becomes so large that we approach the asymptotic value of the line width, which should be, according to the graph, about 16 oersted for a Gaussian. This is in good agreement with the value calculated from the crystal structure of ice (B 15), assuming that the

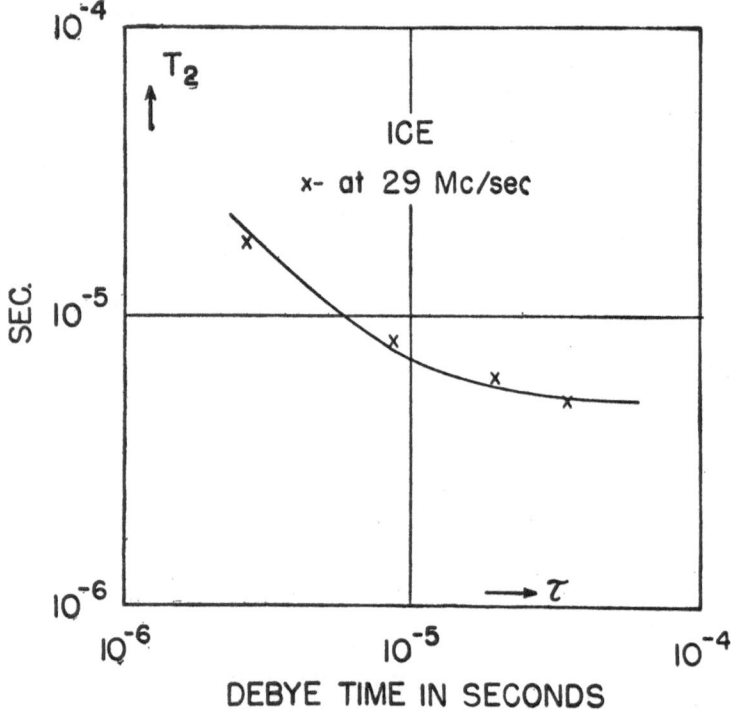

Figure 4. 10.

The line width of the proton resonance in ice between $-2°$ C and $-40°$ C. The theoretical curve (4. 22) for the quantity T_2, which is inversely proportional to the line width, is drawn through the experimental points.

nuclei are at rest. In ice a translational motion of the molecules in a viscous surrounding is apparently excluded. One might assume with D e b y e a hindered rotation of the H_2O molecules in the crystalline structure, although a more recent picture by Onsager suggests, that chains of lined up dipoles will reorient themselves at the positions, where there are misfits with other chains. Either picture will produce the required fluctuations in the local magnetic field and will only affect the proportionality constant between τ and τ_c. The best explanation for the fluctuations in the local field are perhaps the transitions between the two available positions for the proton in the O-H-O bond, as proposed by P a u l i n g (P 8). For comparison the results for alcohol, glycerin and ice at 29 Mc are shown together in fig. 4. 11. For glycerin we can

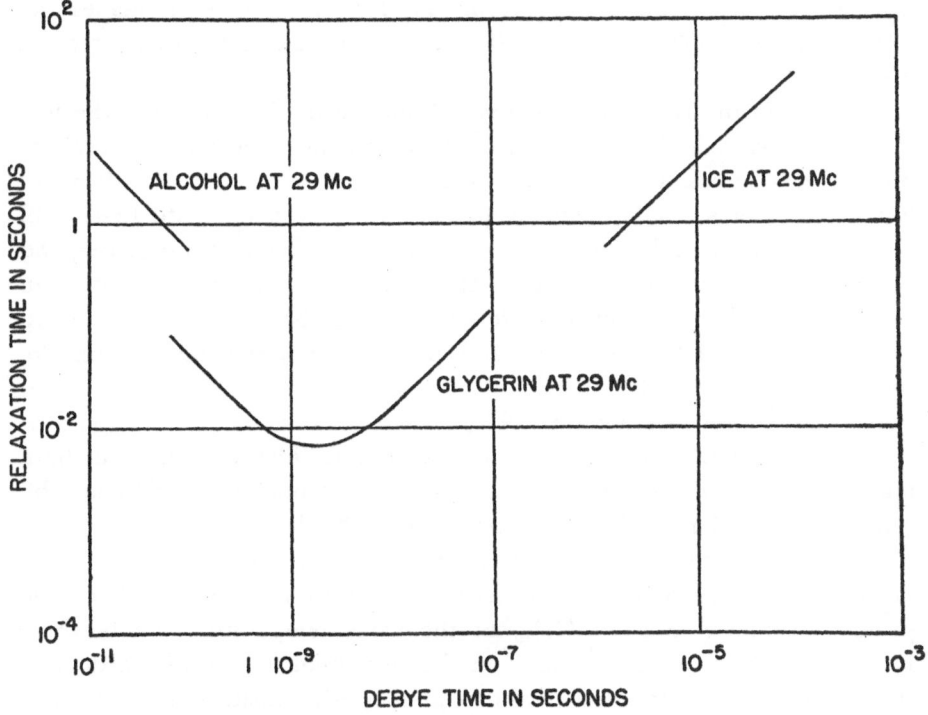

Figure 4. 11.

The relaxation time T_1 of the proton resonance in ethyl alcohol, glycerin and ice at 29 Mc/sec between $-40°$ C and $+60°$ C.

determine the ratio τ/τ_c from comparison of the experimental result of the minimum in the curve with formula (4. 25). We find $\tau_c = 2\,\tau$. Then we must have for alcohol $\tau_c = 0.2\,\tau$ and for ice $\tau_c = 0.8\,\tau$. These results are very satisfactory and must be considered as additional proof for our theory.

We now give a very brief account of what can be expected in other solids with some preliminary experimental results to confirm our view. Much more detailed investigations have to be carried out to refine the following global exposition. In hydrated paramagnetic salts like $CuSO_4$. $5 H_2O$ the field at the position of a proton will fluctuate, because the electron spins of the Cu^{++} ion change their quantisation with respect to H_o at the rate of the short electronic relaxation times ρ, to which we must put equal the correlation time τ_c. The proton resonance in $CuSO_4$. $5 H_2O$ and $CoSO_4$. $7 H_2O$ show line widths of only 12—14 oersted, while the instantaneous value of the internal fields in these paramagnetic salts is several hundred oersted. This can be explained by the short τ_c. The high intensity of local field, arising from the electronic moments, makes the relaxation time so short ($< 3 \times 10^{-4}$ sec), that we could not saturate the proton line.

In paraffin the relaxation time was found to be 0.01 sec. and the line width 4.5 oersted. These data are in agreement with the estimates of other investigators. In molten paraffin the line is narrow. Paraffin behaves again in a similar way as glycerin. In the solid state there still must be an appreciable opportunity for motion, either rotation or twisting or realignment, of the molecules. About the same as for solid paraffin holds for the F^{19} resonance in teflon. This carbon fluoride compound can be considered for our purpose as paraffin, in which the protons are replaced by F^{19} nuclei.

For the proton resonance in NH_4Cl a relaxation time of 0.12 sec. at $+ 20°$ C and 0.015 sec at $- 20°$ C was found. The line width at both temperatures was 4 oersted. These results can probably be explained by a hindered rotation of the NH_4 tetrahedron (S 5).

Very interesting experiments have been carried out by B i t t e r (B 2, A 1), who observed a sharp transition point in the line width of the proton resonance in solid CH_4, at the same temperature where there is known to be a transition point in the rotational degree of freedom of the molecule. The attention of the reader is also called to the measurements at very low temperatures by R o l l i n and collaborators (R 7). Possibly the rotation of the hydrogen molecule can be helpful in explaining the experimental results in solid ortho-hydrogen.

4. 3. 2. *Ionic crystals; the influence of the lattice vibrations.*

4. 3. 2. 1. *The relaxation time.*

We now take up the question of the relaxation time in those crystals, in which lattice vibrations are the only motion. For this case the theory of the relaxation time had been worked out by W a l l e r (W 1, H 2), who considered the interaction of the magnetic moments with the lattice vibrations. We shall show that our procedure, which gave the new results for liquids and gases, is essentially aequivalent to W a l l e r 's considerations, when it is applied to crystals.

For the lattice vibrations we shall adopt the same simplified picture, which D e b y e introduced in his theory of the specific heat of solids (S 5). According to this picture there is an isotropic distribution of lattice oscillators. In the volume V_c of the crystal there are $4 \pi \nu^2 V_c/c^3$ oscillators for one direction of polarisation in the frequency range ν, $\nu + d\nu$.

Here c denotes the velocity of propagation of elastic waves in the crystal, which is taken to be the same for longitudinal and transverse modes.

This formula is valid up to the frequency ν_m determined by the equation

$$\int_0^{\nu_m} 12 \pi \nu^2 V_c c^{-3} d\nu = 3 N \qquad (4.45)$$

For $\nu > \nu_m$ there are no lattice oscillators; (4.45) expresses that the total number of oscillators is equal to the degrees of freedom of the system of N atoms.

We first consider the contribution of one neighbour j to the Fourier spectrum of $\Sigma_j \sin \vartheta_{ij} \cos \vartheta_{ij} e^{i \varphi_{ij}} / r^3_{ij}$.

We take the z-axis in the direction of H_o. The radius vector $\vec{r}_{ij} = \vec{r}_i - \vec{r}_j$ connecting the equilibrium positions of the two nuclei makes an angle ϑ with the z-axis. The displacement $\vec{u}_i$ of the ith nucleus from its equilibrium position by the lattice vibrations is

$$\vec{u}_i = \Sigma_{\nu_k} \vec{A}_k \sin 2 \pi \nu_k (t - r_i/c + \varphi_k) \qquad (4.46)$$

The relative displacement of the ith and j^{th} nucleus for waves propagating in the direction of $\vec{r}_{ij}$

$$\triangle \vec{u}_{ij} = r_{ij} \Sigma_{\nu_k} \frac{2 \pi \nu_k}{c} \vec{A}_k \cos 2 \pi \nu_k (t - r_i/c + \varphi_k) \qquad (4.47)$$

since $\lambda_k = c/\nu_k \gg r_{ij}$. The variation in

$$F_1 = \sin \vartheta_{ij} \cos \vartheta_{ij} e^{i \varphi_{ij}}/r^3_{ij}$$

can be expressed by a Taylor series

$$\triangle F_1 = \frac{\partial F_1}{\partial r} \triangle r + \frac{\partial F_1}{\partial \vartheta} \triangle \vartheta + \frac{\partial F_1}{\partial \varphi} \triangle \varphi + \frac{1}{2} \frac{\partial^2 F_1}{\partial r^2} (\triangle r)^2 + \ldots . (4.48)$$

We dropped the subscripts i and j. For longitudinally polarised waves we have only changes in r; for these $\triangle r = (\triangle u)$ long.

The direction of polarisation of one of the transverse modes is taken in the plane through $\vec{r}_{ij}$ and the z-direction. For this mode we have $r_{ij} \triangle \vartheta = (\triangle u)$ tr. I. For the second transverse mode we have $\mathrm{tg}\triangle \varphi = (\triangle u)$ tr. II$/r \sin \vartheta$. If $\triangle u \ll r \sin \vartheta$, we may write $r \sin \vartheta \triangle \varphi = (\triangle u)$ tr. II. Only for very small ϑ this relation is not satisfied. For this last mode and very small values of ϑ the expansion (4.48) of F is not suitable.

To find the intensity $J_1 (\nu)$ of the spectrum of F_1, we have to determine the sum of the mean square deviations $(\triangle F_1)^2$ in each of the independent waves in the frequency interval ν, $\nu + d\nu$.

We can find an expression for the amplitude A_k of each wave by means of the aequipartition theorem. Each lattice vibrator has an energy $h \nu \left/ \left(e^{\frac{h\nu}{kT}} - 1 \right) \right.$ For small ν or large T this is equal to kT. Let M be the mass of the crystal, $\varrho = M/V_c$ the density. The æquipartition theorem can be written with (4.46) as

$$|A|^2 = \frac{h\nu}{2 \pi^2 \nu^2 M \left(e^{\frac{h\nu}{kT}} - 1 \right)} \approx \frac{kT}{2 \pi^2 \nu^2 M}$$

$$(4.49)$$

We use the last approximation for the three first order terms in (4.48). These terms can be treated independently, as they belong to different directions of polarisation. By squaring each of them and multiplying with the number of oscillators, we find with (4.47, 4.48, 4.49) for the intensity of the spectrum of the first orders terms

$$J_1(\nu) = \frac{4\pi \nu^2 V_c}{3 c^3} \frac{kT}{M} \frac{1}{r^6_{ij}} \left[9 \sin^2 \vartheta_{ij} \cos^2 \vartheta_{ij} + \cos^2 2\vartheta_{ij} + \cos^2 \vartheta_{ij} \right]$$

$$(4.50)$$

A factor $1/3$ is inserted, because the two directions of wave propagation perpendicular to $\vec{r}_{ij}$ do not contribute, as in those waves the two nuclei have the same phase.

Now we can sum (4.50) over all nuclei $j \neq i$. This is legitimate, although there are fixèd phase relations between the deviations of the nuclei in one wave. For the quantisation of the various nuclei is independent, so that their fields aid or counteract at random. If we do not have a single crystal we can average over the angle ϑ, which yields a factor 2 for the expression between brackets in (4.50). The contribution of the Z nearest neighbours at a distance a will be the most important. Applying (2.53) we find for the relaxation time

$$1/T_1 = 4\,\pi\,\gamma^4\,\hbar^2\,I(I+1)\,Z\,k\,T\,\nu_0{}^2/\varrho\,c^5\,a^6 \qquad (4.51)$$

This result is essentially the same as W a l l e r's formula 51 (W 1, p. 386), derived for the transition probability of electronic spins with $I = \frac{1}{2}$. If we take $h\,\nu_0/kT \ll 1$, $\gamma = 2\,\mu/\hbar$ and multiply Waller's result by 2 to get $1/T_1$, we find that our numerical factor is $12\,\pi\,/\,5$ times larger. This difference could probably be explained by noting that W a l l e r used a more detailed picture for the lattice vibrations in a simple cubic lattice. He followed B o r n 's representation of coupled harmonic oscillators. Furthermore W a l l e r quantised the lattice oscillators. To W a l l e r's result and our formula (4.51) a contribution of the processes in which two spins flop simultaneously should be added. It will appear to be much more important, however, to consider the influence of the second order terms in (4.48). On substitution of (4.47) into these terms we see that products of two harmonic functions are present and terms with frequency ν_0 in the expression of ΔF_1 occur as the sum or the difference of two frequencies ν_1 and ν_2. The whole spectrum of the lattice vibrations is important for the second order spectral intensity of F. Since the density of oscillators near the upper limit ν_m is so much higher than at the frequency ν_0, it will turn out that the second order contributions are larger than the first order effects. We find by the same argument which led to (4.50) for the contribution of the first second order term in (4.48) to the spectral intensity

$$J_1{}''(\nu) = \frac{1}{18}\,\sin^2\vartheta_{ij}\cos^2\vartheta_{ij}\,\frac{3^2.4^2}{4\,r^6{}_{ij}\,c^4}\int_{\substack{0 \\ \nu_1\pm\nu_2=\nu_0}}^{\nu_m}\int_0^{\nu_m}\frac{\dfrac{4\,\pi^2\,\nu_1{}^2\,V_c\,h\nu_1}{h\nu_1}}{M c^3\,(e^{\frac{h\nu_1}{kT}}-1)}\cdot\frac{\dfrac{4\,\pi^2\,\nu_2{}^2\,V_c\,h\nu_2}{h\nu_1}}{M c^3\,(e^{\frac{h\nu_1}{kT}}-1)}\,d\nu_1\,d\nu_2$$

and, since $\nu_0 \ll \nu_m$,

$$J_1{}''(\nu) \approx \frac{3\,2\,\pi^2\,h^2}{\varrho^2\,c^{10}}\,\frac{\sin^2\vartheta_{ij}\cos^2\vartheta_{ij}}{r^6{}_{ij}}\int_0^{\nu_m}\frac{\nu'^6}{(e^{\frac{h\nu'}{kT}}-1)^2}\,d\nu' \qquad (4.52)$$

Since all frequencies up to ν_m are involved, we cannot make use of of the condition $x = h\,\nu'/k\,T \ll 1$, unless the temperature T is large compared to the Debye temperature $\Theta = h\,\nu_m/k$ of the crystal. The relaxation time, determined by this second order process, is by the same arguments which led to (4.51),

$$1/T_1 = \frac{8}{5}\,\gamma^4\,Z\,I(I+1)\,\frac{(k\,T)^7}{\varrho^2\,c^{10}\,a^6\,h^8}\int_0^{\Theta/T}\frac{x^6}{(e^x-1)^2}\,d\,x \qquad (4.53)$$

To (4.53) should be added the result of the other second order terms and the contribution of the double processes, in which two spins make a simultaneous transition. The numerical factor in (4.53) would become somewhat larger. But as it is, it is already $18\,\pi^2 \times 192/245$ larger than in W a l l e r's formula 56 (p. 388) for the quantised lattice oscillators. In the language of quantummechanics we can say that to (4.53) correspond transitions of the nuclear spin accompanied by the emission of a phonon and the absorption of another in the lattice. One could develop (4.48) to the third order terms, etc. It turns out that the contribution of the successive higher terms decreases as $k\,T\,\nu_m^3/\varrho\,c^5 \approx 10^{-2}$; so they can be neglected.

We see from (4.51) and (4.53) that the first order transition probability goes as T, the second order one as T^2 for $\Theta/T \approx 1$ but as T^7 for $\Theta/T \gg 1$. At room temperature the second order terms are more important. Substituting numerical values $\varrho = 2$, $c = 2\times10^5$, $\nu_0 = 3\times10^7$, $a = 2\times10^{-8}$, $Z = 6$, $\gamma = 3\times10^4$, $T = 300°$, $\Theta \ll T$ we find that (T_1) first order $\approx 10^{14}$ sec and (T_1) second order $\approx 10^3$ sec.

It was a surprise that, while W a l l e r's theory predicted such long relaxation times for the nuclear magnetic resonance, the first experimental results gave much shorter times (10^{-2} sec in paraffin). We have shown that in many solids the spectral intensity of the local field is caused by other motions than the lattice vibrations and that so many observed relaxation times could be explained. In ionic crystals like Ca F_2, however, one would expect W a l l e r's theory to be applicable. Nevertheless the relaxation time for the F^{19} resonance in a single crystal of Ca F_2 appeared to be 8 sec. Relaxation times of the order of one second were also found in powdered Al F_3 and Na F, and by other authors in Li F. There are some indications that impurities and lattice defects play an important role in the relaxation process of these crystals (Compare the note at the end of this chapter).

4. 3. 2. 2.   *The line width.*

The line width must be calculated from the components near zero frequency in the spectrum of $F_0 = \sum_{ij} (1 - 3 \cos^2 \vartheta_{ij})/r^3_{ij}$. In the evaluation we can safely neglect the small and rapid lattice vibrations and assume that the nuclei are at rest. For this static problem the line width is given by (2. 36). It should be independent of the temperature, but vary with the orientation of the axes of a single crystal with respect to the direction of $\vec{H_o}$. Experiments (P 5) with a single crystal of Ca F_2 gave results for the line width in accordance with (2. 36) applied to the simple cubic lattice of F^{19} nuclei, the Ca ions having no magnetic moment. A detailed investigation of the line width in solids with special attention to the line shape was made by P a k e (P 1). In many compounds the same element can occur in more than one position in the unit cell of the crystal. When these positions are not aequivalent with respect to the internal magnetic field, one should distinguish more than one relaxation time and line width at the resonance of those nuclei. It is of no use, however, to discuss the situation in crystalline solids in detail, before more experimental material has become available.

Note added in the proof:

Recents experiments carried out in the Kamerlingh Onnes Laboratory of the University of Leiden confirm the hypothesis that the relaxation mechanism in ionic crystals is determined by paramagnetic impurities.

A theory, taking these into account, gives for T_1 a value of the order of a few seconds, if the crystal is contaminated with 0.0001 % iron. Furthermore this theory predicts that T_1 should be largely independent of the temperature of the lattice. These features are in striking contrast with W a l l e r's results for an ideal lattice and agree much better with the experimental data (comp. R 7).

A full account of these researches will be given elsewhere.

CHAPTER 5

RELAXATION BY QUADRUPOLE COUPLING.

5. 1. *The influence of the quadrupole moment on the relaxation time and line width.*

If a nucleus has a spin $I > \frac{1}{2}$, the possibility of a spherically asymmetrical charge distribution over the nucleus exists. The value of the electric quadrupole momente $e\,Q$ is defined in C a s i m i r's basic treatise (C 2) on the quantummechanics of the quadrupole moment as (compare however B 14)

$$e\,Q = \int (3\,z^2 - r^2)\,\varrho_{m_I = I}(r)\,d\,\tau \qquad (5.\,1)$$

The integral is extended over the volume of the nucleus. The nuclear charge distribution is taken for the state with maximum z-component of the angular momentum $m_I = I$. The quantummechanical operator associated with the electric quadrupole moment consists of the components of a symmetrical tensor, the diagonal sum of which is zero. Since the group of rotations transforms these components in the same way as the Legendre polynomia of the second order, the components of the quadrupole moment can be shown to have the form

$$Q_{k\,l\,op.} = \frac{e\,Q}{I\,(2\,I - 1)} \left[\frac{3}{2}\,(I_k\,I_l + I_k\,I_l) - \delta_{k\,l}\,I^2 \right]_{op.} \qquad (5.\,2)$$

The subscripts k, l, can each denote the x-, y- or z-component. Each of the components of the tensor is a matrix over the magnetic quantumnumber m_I. The matrix elements can easily be evaluated from (5. 2) with the given form (1. 1), (1. 2) and (1. 3) of the operators I_x, I_y and I_z. It is directly seen that $Q_{k,l}$ has only elements connecting states with $\Delta m = \pm 2, \pm 1$ or 0.

When a quadrupole moment is present, we have to add a term to the Hamiltonian

$$H_{Q\,op.} = \sum_{\text{all nuclei}} \left[-\frac{1}{6} \sum_k \sum_l Q_{k\,l} \times \frac{1}{2} \left(\frac{\partial_l}{\partial E_k} + \frac{\partial_k}{\partial E_l} \right) \right] \qquad (5.\,3)$$

The gradient of the components of the electric field $\vec{E}$, which occurs in this expression, is, of course, entirely due to the charge distribution in the sample. We do not apply an external inhomogeneous electric field. In heavy water e.g. the inhomogenous field at the position of a deuteron arises in the first place from the asymmetrical charge distribution of the other constituents in the same water molecule, but also from the electric dipoles of the neighbouring water molecules. We can now apply to the electric quadrupole perturbation (5.3) the same considerations as we did in chapter 2 to the magnetic dipole interaction (2.33). We note that because of the thermal motion of the molecules in the sample we again have a frequency spectrum of each of the components of grad $\vec{E}$. The correlation time τ_c for these components will be about the same as for the magnetic field, since for both τ_c is the time in which the position of the molecules with respect to H_o and one another has changed appreciably. The intensity of the spectrum of grad $\vec{E}$ at the Larmor frequency ν_0 of the nuclei is responsible for quadrupole transitions with $\Delta m = \pm 1$, the intensity of the spectrum at $2\nu_0$ for quadrupole transitions with $\Delta m = \pm 2$. These transitions shorten the relaxation time T_1. The components near the frequency zero in combination with the diagonal matrix elements of Q_{op} will broaden the resonance line. As an example of the terms we have to add to the formulae (2.51) and (2.53) for the relaxation time and line width, we write down the contribution to T_1 by the quadrupole moment of a nucleus with $I = 1$:

$$(1/T_1)_Q = \frac{1}{4}\frac{e^2 Q^2}{\hbar^2}\frac{\tau_c}{1 + 16\,\pi^2\nu^2\,\tau_c^2}\left\{\left(\frac{\partial E_x}{\partial x}\right)^2 + \left(\frac{\partial E_y}{\partial y}\right)^2 + \frac{1}{4}\left(\frac{\partial E_x}{\partial y} + \frac{\partial E_y}{\partial x}\right)^2\right\} +$$

$$+ \frac{1}{16}\frac{e^2 Q^2}{\hbar^2}\frac{\tau_c}{1 + 4\,\pi^2\nu^2\,\tau_c^2}\left\{\frac{1}{4}\left(\frac{\partial E_x}{\partial z} + \frac{\partial E_z}{\partial x}\right)^2 + \frac{1}{4}\left(\frac{\partial E_y}{\partial z} + \frac{\partial E_z}{\partial y}\right)^2\right\} \qquad (5.4)$$

5.2. Experimental results for the resonance of H^2 and Li^7.

The influence of the quadrupole moment on the relaxation time has been observed for the resonance of the deuteron in water. Two samples were used. One contained 0.4 cc of a mixture of normal and heavy water, the other consisted of the same mixture with some $CuSO_4$ added. In both samples 51 % of the hydrogen nuclei were deuterons. The magnetic resonance of both protons and deuterons was observed at 4.8 Mc/sec and saturation curves were taken which are shown in fig. 5.1. For comparison the proton resonance in normal water was also measured.

The relaxation time for the proton resonance in $H_2O + D_2O$ is about 1.4 times longer than in H_2O. This is due to the decreased intensity of the local magnetic field, since half of the protons are replaced by deuterons, of which the magnetic moment is about three times smaller.

The intensity of the local field at the position of a deuteron will be about the same as for a proton. It will be slightly higher because the percentages of H_2O, HDO and D_2O molecules in the liquid are such that the chance of a nearest neighbour of a deuteron to be a proton is

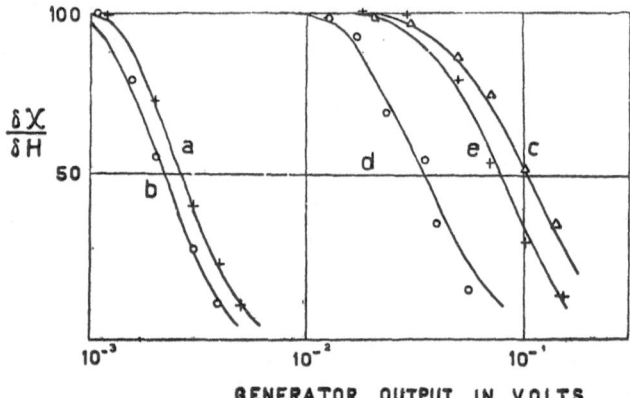

Figure 5. 1.

The saturation of the proton and deuteron resonance at 4.8 Mc/sec in light and heavy water.
a) Proton resonance in H_2O;
b) Proton resonance in 49 % H_2O + 51 % D_2O;
c) Proton resonance in 49 % H_2O + 51 % D_2O + $CuSO_4$;
d) Deuteron resonance in 49 % H_2O + 51 % D_2O;
e) Deuteron resonance in 49 % H_2O + 51 % D_2O + $CuSO_4$.
The relaxation times calculated from the saturation curves can be found in the text.

larger than 49 %. The local field intensity, however, can certainly not be higher than in pure H_2O.

Suppose for the moment, that the deuteron would have no quadrupole moment. Then we expect, that the saturation of the d-resonance would occur at the same energy density of the applied radiofrequency field as the proton resonance, namely, when this density becomes comparable to that of the local field. We see, however, from fig. 5. 1 that the energy density, required for the saturation of the d-resonance, is 180 times higher. Taking into account that $\gamma^2_p/\gamma^2_d = 39$, we find that the relaxation time for the d-resonance in water is 0.5 sec, while it should be 90 sec, if no quadrupole moment were present. The explanation must be that the quadrupole moment of the deuteron is almost solely responsible for the observed short relaxation time. Substituting the values for T_1 and $Q = 2.73 \times 10^{-27}$ cm² in (5.4) we can estimate the value of grad $\vec{E}$ in the liquid. It is the same as would be produced by one elementary charge at a distance of one Ångstrøm from the deuteron. This is a reasonable value for the inhomogeneity of the electric field in the molecule, surrounded by electric dipoles. So even small quadrupole moments can have a considerable effect.

Small concentrations of a paramagnetic salt will have no effect on the relaxation time, as the electric transitions remain more important at first than the magnetic ones. But the addition of a sufficient amount of $CuSO_4$ will increase the magnetic local field density so much that the relaxation time is then determined only by the magnetic transitions. The saturation of the p- and d-resonance then occurs at the same value of the applied field. The curves c and e show this situation. They should coincide exactly. The deviation of a factor 1.8 is probably due to a systematic error (cf. section 3.5). The inhomogeneity of the magnetic field for the p-resonance (at about 1100 oersted) is certainly different from that for the d-resonance (at about 7000 oersted).

One might ask why the influence of the quadrupole moment on the relaxation time of Li^7 is not more pronounced. In the first place is the magnetic moment of Li^7 rather large, making the relative influence of Q smaller.

Let us assume that the relaxation time of Li^7 in an aqueous solution, which according to tabel IV is 1.75 sec, is for 50 % due to quadrupole interactions. The quadrupole moment Q of Li^7 is not known experimentally. A very rough theoretical estimate by W e l l e s (W 4), built on the Hartree model, gives a value of -2.7×10^{-26} cm². On substitution of this value and $T_1 = 3.5$ sec in (5.4) we find that the gradient of the electric field is the same as produced by an elementary charge at least 3 Ångstrøm away. This distance seems to be too large, although one must expect that grad E is much smaller than in the case of the deuteron, since the Li^+ ion is in a ^1S state and the neighbouring water dipoles will arrange themselves around the ion so as to give an approximately spherical charge distribution. It seems probable, that the quadrupole moment of Li^7 is about five times smaller than the theoretical estimate, mentioned above.

The line width of the D^2 and Li^7 resonance in the liquids used is determined by the inhomogeneity of the field H_o. The influence of the quadrupole moment could not be detected. P o u n d (P 9) recently has found broad resonance lines of the two Br isotopes in a solution of NaBr.

The width must be ascribed to the large quadrupole moments of these isotopes. P o u n d also discovered a fine structure of the Li[7] resonance in a single crystal of $Li_2 SO_4$. The diagonal elements of (5.3), in which grad $\vec{E}$ can now be considered as a constant, give the first order perturbation of the four levels m_I which the Li[7] nucleus can occupy in a magnetic field. If grad $\vec{E}$ is known from the crystal structure, the observed shift of the resonance frequencies enables one to determine the quadrupole moment Q.

5.3. *The quadrupole interaction in free molecules.*

The quadrupole interaction can be described more precisely, if we have to do with only one molecule. This is the case in molecular beam experiments (K 4) and in the application of the theory of the relaxation time to deuterium gas.

The grad $\vec{E}$ can then be written in terms of the angular momentum $\vec{J}_{op}$ of the molecule. By grouptheoretical arguments we have

$$\frac{1}{2}\left(\frac{\partial E_k}{\partial l} + \frac{\partial E_l}{\partial k}\right) = -\frac{eq}{J(2J-1)}\left[\frac{3}{2}\left(J_k J_l + J_l J_k\right) - \delta_{kl} J^2\right]_{op} \quad (5.5)$$

$$\text{with } eq = \int \frac{3z^2 - r^2}{r^5}\, \varrho_{m_J = J}(\mathfrak{r})\, d\tau \quad (5.6)$$

The integration has to be extended over the charges of the molecule outside the nucleus under consideration, and for the state $m_J = J$.

The interaction term in the Hamiltonian can then be written as

$$H_{Q_{op}} = \frac{1}{2}\frac{e^2 qQ}{I(2I-1)J(2J-1)}\left[3\left(\vec{I}.\vec{J}\right)^2 + \frac{3}{2}\left(\vec{I}.\vec{J}\right) - \vec{I}^2 \vec{J}^2\right]_{op} \quad (5.7)$$

We note that this operator has the same form as (4.36) and can be dealt with in the same way as described in chapter 4. For D_2 gas we must expect a shorter relaxation time than in H_2 gas, for the quadrupole interaction in the D_2 molecule is much larger than the magnetic interaction in H_2, as follows from the splitting of the lines of the resonance spectra obtained with the molecular beam method.

Finally we must mention a refinement of the theory given in chapter 2. When we consider the dipole-dipole interaction between two identical protons in the same molecule, we should distinguish between ortho- and para-states, as we did for the hydrogen molecule. The magnetic interaction in a linear molecule will again assume the form (4.36). In a liquid J will probably not be constant of the motion which would make the problem involved. But J will usually be very high. The very light molecules like H_2 and D_2 are an exception in this respect. When J is large, we can introduce the classical approximation and replace J_z/J by $\cos \vartheta$ and $J_x + iJ_y/J$ by $\sin \vartheta\, e^{i\varphi}$ in the matrix elements (4.37). So we come back to formulae of the same form as (2.34) previously derived.

We do not touch the question how exchanges between nuclei in more complicated nuclei and between nuclei in neighbouring molecules should be taken into account. Under many circumstances — but not in H_2 and D_2 — the formulae (2.51) and (5.4), where all nuclei have been considered as distinguishable and the motion of the molecules is not quantised, will give a satisfactory description.

SUMMARY.

The exchange of energy between a system of nuclear spins immersed in a strong magnetic field, and the heat reservoir consisting of the other degrees of freedom (the "lattice") of the substance containing the magnetic nuclei, serves to bring the spin system into equilibrium at a finite temperature. In this condition the system can absorb energy from an applied radiofrequency field. With the absorption of energy however, the spin temperature tends to rise and the rate of absorption to decrease. Through this "saturation" effect, and in some cases by a more direct method, the *spin-lattice relaxation time* T_1 can be measured. The interaction *among* the magnetic nuclei, with which a characteristic time T_2' is associated, contributes to the width of the absorption line. Both interactions have been studied in a variety of substances, but with the emphasis on liquids containing hydrogen.

Magnetic resonance absorption is observed by means of a radiofrequency bridge; the magnetic field at the sample is modulated at a low frequency. A detailed analysis of the method by which T_1 is derived from saturation experiments is given. Special attention is paid to the influence of the inhomogeneity of the external magnetic field and to the limitation of the accuracy by noise. Relaxation times observed range from 10^{-4} to 10 seconds. In liquids T_1 ordinarily decreases with increasing viscosity, in some cases reaching a minimum value after which it increases with further increase in viscosity. The line width meanwhile increases monotonically from an extremely small value toward a value determined by the spin-spin interaction in the rigid lattice. The effect of paramagnetic ions in solution upon the proton relaxation time and line width has been investigated. The relaxation time and line width in ice have been measured at various temperatures.

The results can be explained by a theory which takes into account the effect of the thermal motion of the magnetic nuclei upon the spin-spin interaction. The local magnetic field produced at one nucleus by neighbouring magnetic nuclei, or even by electronic magnetic moments of paramagnetic ions, is spread out into a spectrum extending to frequencies of the order of $1/\tau_c$] where τ_c is a correlation time associated with the local Brownian motion and closely related to the characteristic time which occurs in D e b y e's theory of polar liquids. If the nuclear Larmor frequency ω is much less than $1/\tau_c$, the perturbations due to the local field nearly average out, $T_1 \sim 1/\tau_c$, and the width of the resonance line, in frequency, is about $1/T_1$. A similar situation is found in hydrogen gas where τ_c is the time between collisions. In very viscous liquids and in some solids where $\omega\,\tau_c > 1$, a quite different behavior is predicted, and observed.

Values of τ_c for ice, inferred from nuclear relaxation measurements, correlate well with dielectric dispersion data.

When the theory is applied to the motion embodied by the lattice vibrations of a crystal, it becomes identical to that of W a l l e r. The values for T_1 predicted by this theory are several orders of magnitude larger than those observed experimentally in ionic crystals.

The theory is also extended to the interaction of an electric quadrupole moment with an inhomogeneous internal electric field. The results are in good agreement with the observed relaxation time for the D^2-resonance in heavy water.

REFERENCES

A 1. N. L. Alpert, Phys. Rev. **72**, 637, 1947.

A 2. L. W. Alvarez and F. Bloch, Phys. Rev. **57**, 111, 1940.

A 3. H. L. Anderson and A. Novick, Phys. Rev. **71**, 878, 1947.

A 4. W. R. Arnold and A. Roberts, Phys. Rev. **71**, 878, 1947.

A 5. H. L. Anderson and A. Novick, Phys. Rev. (in press).

B 1. H. A. Bethe and R. F. Bacher, Rev. Mod. Phys. **8**, 82, 1936.

B 2. F. Bitter, N. L. Alpert, H. L. Poss, C. G. Lehr and S. T. Lin, Phys. Rev. **71**, 738, 1947.

B 3. P. Blacket and F. Champion, Proc. Roy. Soc. **A 130**, 380, 1930.

B 4. F. Bloch, Phys. Rev. **70**, 460, 1946.

B 5. F. Bloch, A. C. Graves, M. Packard and R. W. Spence, Phys. Rev. **71**, 373, 1947; **71**, 551, 1947.

B 6. F. Bloch, W. W. Hansen and M. Packard, Phys. Rev. **69**, 127, 1946.

B 7. F. Bloch, W. W. Hansen and M. Packard, Phys. Rev. **70**, 474, 1946.

B 8. F. Bloch and I. I. Rabi, Rev. Mod. Phys. **17**, 237, 1946.

B 9. F. Bloch and A. Siegert, Phys. Rev. **57**, 522, 1940.

B 10. N. Bloembergen, R. V. Pound and E. M. Purcell, Phys. Rev. **71**, 466, 1947.

B 11. N. Bloembergen, E. M. Purcell and R. V. Pound, Nature **160**, 475, 1947.

B 12. L. J. F. Broer, *Beschouwingen en metingen over de paramagnetische relaxatie,* Thesis, Amsterdam 1945.

B 13. L. J. F. Broer, Physica **10**, 801, 1943.

B 14. J. Bardeen and C. H. Townes, Phys. Rev. **73**, 97, 1948.

B 15. J. O. Bernal and R. H. Fowler, J. Chem. Phys. **1**, 515, 1933.

C 1. H. B. G. Casimir, Physica **7**, 169, 1940.

C 2. H. B. G. Casimir, *On the interaction between atomic nuclei and electrons,* Archives du Musee Teyler, **8**, 202, 1936.

C 3. D. E. Coles and W. E. Good, Phys. Rev. **70**, 979, 1946.

C 4. E. H. Condon and G. H. Shortley, *Theory of atomic spectra,* Cambridge 1935.

C 5. C. H. Collie, J. B. Halsted and D. M. Ritson, Proc. Phys. Soc. **60**, 71, 1948.

C 6. R. L. Cummerow, D. Halliday and G. E. Moore, Phys. Rev. **72**, 1233, 1947.

D 1. B. P. Dailey, R. L. Kyhl, M. W. P. Strandberg, J. H. van Vleck and E. B. Wilson Jr., Phys. Rev. **70**, 984, 1946.

D 2. P. Debye, *Polar Molecules,* New York 1945, chapter 5.

D 3. R. Dicke, Rev. Sci. Inst. **17**, 268, 1946.

E 1. A. E i n s t e i n, Ann. Phys. **17**, 549, 1905; **19**, 371, 1906.

E 2. I. E s t e r m a n n and O. S t e r n, Z. Phys. **85**, 17, 1933.

F 1. A. F a r k a s, *Light and Heavy Hydrogen,* Cambridge 1935.

F 2. R. F r i s c h and O. S t e r n, Z. P h y s. **85**, 4, 1933.

G 1. C h r. G e h r t s e n, Phys. Z. **38**, 833, 1937.

G 2. E. G e r j u o y and J. S c h w i n g e r Phys. Rev. **61**, 138, 1942.

G 3. C. J. G o r t e r, *Paramagnetic relaxation,* Amsterdam-New York, 1947.

G 4. C. J. G o r t e r, Physica **3**, 995, 1936.

G 5. C. J. G o r t e r and L. J. F. B r o e r, Physica **9**, 591, 1942.

G 6. C. J. G o r t e r and R. K r o n i g, Physica **3**, 1009, 1936.

G 7. P. G u t t i n g e r, Z. Phys. **73**, 169, 1931.

H 1. W. H e i t l e r, *Quantumtheory of radiation,* Oxford 1936.

H 2. W. H e i t l e r and E. T e l l e r, Proc. Roy. Soc. **A 155**, 629, 1936.

J 1. L. C. J a c k s o n, Proc. Phys. Soc. **50**, 707, 1938.

J 2. G. J o o s, *Lehrbuch der Theoretischen Physik,* Leipzig 1939, page 205.

K 1. K. J. K e l l e r, Thesis, Utrecht 1943.

K 2. J. M. B. K e l l o g g and S. M i l l m a n, Rev. Mod. Phys. **18**, 323, 1946.

K 3. J. M. B. K e l l o g g, I. I. R a b i, N. F. R a m s e y and J. R. Z a c h a r i a s, Phys. Rev. **56**, 728, 1939.

K 4. J. M. B. K e l l o g g, I. I. R a b i, N. F. R a m s e y and J. R. Z a c h a r i a s, Phys. Rev. **57**, 677, 1940.

K 5. H. K o p f e r m a n n, *Kernmomente,* Leipzig 1940.

K 6. H. A. K r a m e r s, Hand- und Jahrbuch der Chem. Physik I, Leipzig 1937.

K 7. H. A. K r a m e r s, Atti Congr. Fis. Como, 545, 1927.

K 8. R. K r o n i g, Physica **5**, 65, 1938.

K 9. R. K r o n i g, J.O.S.A., **12**, 547, 1926.

K 10. R. K r o n i g and C. J. B o u w k a m p, Physica **5**, 521, 1938.

K 11. R. K i n g, H. R. M i m n o and H. W i n g, *Transmission lines, wave guides and antennae,* New York 1942.

K 12. P. K u s c h, S. M i l l m a n and I. I. R a b i, Phys. Rev. **55**, 1176, 1939.

L 1. B. L a s a r e v and L. S h u b n i k o w, Phys. Z. Sowjetunion, **11**, 445, 1937.

M 1. E. M a j o r a n a, Nuovo Cimento, **9**, 43, 1932.

M 2. J. M. W. M i l a t z and N. B l o e m b e r g e n, Physica **11**, 449, 1945.

M 3. S. M i l l m a n, Phys. Rev. **55**, 628, 1939.

M 4. N. F. M o t t, Proc. Roy. Soc. **A 126**, 259, 1930.

P 1. G. E. P a k e, J. Chem. Phys. (in press).

P 2. W. P a u l i, Naturwissenschaften **12**, 741, 1924.

P 3. R. V. P o u n d, E. M. P u r c e l l and H. C. T o r r e y, Phys. Rev. **69**, 680, 1946.

P 4. E. M. P u r c e l l, Phys. Rev. **69**, 681, 1946.

P 5. E. M. P u r c e l l, N. B l o e m b e r g e n and R. V. P o u n d, Phys. Rev. **70**, 988, 1946.

P 6. E. M. P u r c e l l, R. V. P o u n d and N. B l o e m b e r g e n, Phys. Rev. **70**, 986, 1946.

P 7. E. M. P u r c e l l, H. C. T o r r e y and R. V. P o u n d, Phys. Rev. **69**, 37, 1946.

P 8. L. P a u l i n g, *The Nature of the Chemical Bond,* p. 302 ff, Cornell 1940.

P 9. R. V. P o u n d, Phys. Rev. **72**, 1273, 1947.

R 1. I. I. R a b i, Phys. Rev. **51**, 652, 1937.

R 2. I. I. R a b i, S. M i l l m a n, P. K u s c h and J. R. Z a c h a r i a s, Phys. Rev. **55**, 526, 1939.

R 3. W. Rarita and J. Schwinger, Phys. Rev. **59**, 436 and 556, 1941.

R 4. S. O. Rice, Bell Tel. J. **23**, 282, 1944; **25**, 46, 1945.

R 5. B. V. Rollin, Nature **158**, 669, 1946.

R 6. B. V. Rollin and J. Hatton, Nature **159**, 201, 1947.

R 7. B. V. Rollin, J. Hatton, A. H. Cooke and R. J. Benzie, Nature **160**, 436, 1947.

R 8. M. E. Rose, Phys. Rev. **53**, 715, 1938.

S 1. R. Sachs, Phys. Rev. **71**, 457, 1947.

S 2. H. Schüler and Th. Schmidt, Z. Phys. **94**, 457, 1935.

S 3. J. Schwinger, Phys. Rev. **51**, 648, 1937.

S 4. J. Schwinger, Phys. Rev. **55**, 235, 1939.

S 5. F. Seitz, *Modern Theory of Solids,* New York 1940.

T 1. C. H. Townes, A. N. Holden, J. Bardeen and F. R. Merritt, Phys. Rev. **71**, 644, 1947.

V 0. G. A. Valley and H. Wallman, Radiation Laboratory Series, Vol. 18, Mc Graw-Hill.

V 1. J. H. van Vleck, *Theory of electric and magnetic susceptibilities,* Oxford 1932.

V 2. J. H. van Vleck, Phys. Rev. **55**, 924. 1939.

V 3. J. H. van Vleck, Phys. Rev. (in press).

W 1. I. Waller, Z. Phys. **79**, 370, 1932.

W 2. Ming Chen Wang and G. E. Uhlenbeck, Rev. Mod. Phys. **17**, 323, 1945.

W 3. V. Weiskopf and E. Wigner, Z. Phys. **63**, 54, 1930; **65**, 18, 1930.

W 4. S. H. Welles, Phys. Rev. **62**, 197, 1942.

W 5. H. Wintsch, Helvetica Phys. Acta **5**, 126, 1932.

Z 1. E. Zavoisky, J. Phys. U.S.S.R **9**, 211, 1945; ibid. 245 and 447.

ERRATA

page

17   formula 1. 13, for „$w_{m \leftarrow m'}$" read: „$w_{m' \leftarrow m}$".
23   4th line from top, for „aud" read: „and".
27   3th line from bottom, for „is" read: „if".
31   5th line from top, add „a" between „has" and „sharp".
31   formula 2. 26, for „ϱ_q" read: „ϱ_p".
32   18th line from bottom, omit „the".
37   11th line from top, for „distingiush" read: „distinguish".
39   formula 2. 44 should read:

$$\frac{d N^+}{d t} = - W N^+ exp \frac{\gamma \hbar H_0}{k T} + W N^- exp - \frac{\gamma \hbar H_0}{k T}.$$

42   8th line from top, for „to" read: „two".
46   10th line from top, read:

„ . . . . . for $T_2 \triangle \omega = \sqrt{1 + \gamma^2 H_1^2 T_1 T_2}$ and . . . . .".

50   formula 2. 77, the lower limit of the integral should be „$-\infty$".
58   4th line from bottom, add reference „(V 0)".
108 The value of γ for He^8 is not 2.4×10^4.
     For correct value see reference (A 5).